DRAWING (...) CITY (...) BODY, DWELLING ON EARTH

PROCEEDINGS OF THE 5TH INTERNATIONAL SEMINAR ARCHITECTURAL GRAPHIC REPRESENTATION AND OTHER IMAGES, SÃO PAULO, BRAZIL, 5–9 MARCH 2018

Drawing (...) City (...) Body, Dwelling on Earth

Imagined-Architectures: Architectural graphic representation and other images

Editor

Pedro António Janeiro
CIAUD – Research Centre for Architecture, Urbanism and Design, University of Lisbon, Portugal

CRC Press
Taylor & Francis Group
Boca Raton London New York

CRC Press is an imprint of the
Taylor & Francis Group, an **informa** business
A BALKEMA BOOK

CRC Press/Balkema
P.O. Box 447, 2300 AK Leiden, The Netherlands
e-mail: Pub.NL@taylorandfrancis.com
www.crcpress.com – www.taylorandfrancis.com

ISBN 13: 978-0-367-02602-8 (hbk)
ISBN 13: 978-0-429-39877-3 (ebk)

Visit the Taylor & Francis Web site at
http://www.taylorandfrancis.com

and the CRC Press Web site at
http//www.crcpress.com

Typeset by V Publishing Solutions Pvt Ltd., Chennai, India

Drawing (...) City (...) Body, Dwelling on Earth – Janeiro (Ed.)
© 2019 Taylor & Francis Group, London, ISBN 978-0-367-02602-8

Table of contents

Preface vii

Committees ix

Art and Morphology: The legacy of Camillo Sitte 1
A. Barbosa & R. Duque

The drawing of the winding line 7
A. Godinho

Former "Portuguese" Sub-Saharan colonial Africa: A drawing narrative 15
A. Milheiro

Drawing as a key feature in architectural creative process 21
A. Moreira da Silva

The thickness of the world 27
D. Loução & M. Cruz

(Para)metric representations and urban indicators: An application of citymetrics in a neighborhood of João Pessoa, Brazil 37
F. Lima & G. Silva

Hundertwasser five skins: Pretext for the study of the role of drawing and human body in the project of architecture 45
F. Moreira da Silva

The mask of the classical theater for a Tactile Museum, through artisanal approaches and digital protocols 51
F. Fatta & A. Marraffa

Cinema's places and territories: A systematic catalogue of Italian movie theaters 65
F. Moschini, F. Maggiore & V. D'Alba

The city skin – visions for a *tactile* identity of the space 79
F. Oliveira

The classical composition in the social spaces of housing in Recife, Brazil 87
G. Carvalho, I. Bechmann & M. Figueiredo

Walking: Drawing the city with the body 95
J. Ferreira Crespo

The comic as an academic product: Presence and territoriality 101
C.G. Abdalla & C. Santos

Urban vitality in Madureira: Field observation and recording 111
J. Rocha & M. Costa

City Portrait: A pedagogical experience in the Faculty of Architecture and Urbanism of University of São Paulo 121
L. Jorge

Legal alien, exile and intermittent homes in "Englishman in New York" (Sting, 1987) 131
L. Sol

Dwelling on earth… about the Kampung of Tongkol in Jakarta 137
M. Louro & V. Catalfamo

Body in the space: Eduardo Gageiro's "drawing" photography 147
M. Baptista-Bastos

Insightful drawings; Invented cities; Inverted bodies 155
M. Nascimento & F. Khoury

A contra hegemonic design for the Brazilian cities, the Rio de Janeiro case 163
P. da Luz Moreira

Possibilities of the flesh of the body and the flesh of the things: The drawing as a conversion of the *visible into an other-visible* 173
P. Janeiro

Possibilities of overcoming aesthetic preconceptions in drawing learning: Exercise notes about sensitive impressions 179
R. Ferreira Lopes & L. Mendes de Vasconcellos

Scarpa in Sicily: The memory of our senses 189
C. Giunta

Literary and travel images for the knowledge of the city 203
V. Cardone, S. D'Auria & E. De Feo

Conversions 213
E.W. Angelico

Author index 219

Drawing (...) City (...) Body, Dwelling on Earth – Janeiro (Ed.)
© 2019 Taylor & Francis Group, London, ISBN 978-0-367-02602-8

Preface

The Research Project *Arquitecturas-Imaginadas: Representação Gráfica Arquitectónica e 'Outras-Imagens'* (Imagined Architectures: Architectural Graphic Representation and Other Images), based since 2009 at the Research Centre for Architecture, Urbanism and Design, at the Lisbon School of Architecture, University of Lisbon, C.I.A.U.D./F.A./U.Lisboa, held in partnership with the University of São Paulo and with the Federal University of Juiz de Fora, Brazil, its fifth International Seminar, dedicated to the theme *Desenho (...) Cidade (...) Corpo, Habitando a Terra* (Drawing [...] City [...] Body, Inhabiting the Earth).

This International Seminar was held as a joint activity between: this C.I.A.U.D./F.A./U. Lisboa Research Project, the University of São Paulo, represented by the Maria Antônia University Centre, and the Faculty of Architecture and Urbanism of the Federal University of Juiz de Fora; having as scientific heads Professor Pedro António Janeiro (C.I.A.U.D./ F.A./U.Lisboa), Myrna de Arruda Nascimento (FAU/USP – Maria Antônia University Centre), and Ricardo Ferreira Lopes (DPRT/FAU/UFJF), respectively.

In the section on Presentation and Regulations, at the time of the Call for Papers for this International Seminar, the following text was provided:

> In sum, this International Seminar's fifth edition dedicated to the theme *Desenho (...) Cidade (...) Corpo, Habitando a Terra* (Drawing [...] City [...] Body, Inhabiting the Earth) – similarly to its first edition, entitled *Desenho (...) Cidade* (Drawing [...] City), of April 2014; similarly to its second edition, entitled *Desenho (...) Cidade (...) Eu* (Drawing [...] City [...] Myself), of April 2015, both held at the Lisbon School of Architecture, University of Lisbon; similarly to its third edition, entitled *Desenho (...) Cidade-Moderna* (Drawing [...] Modern City), held at the Faculty of Architecture and Urbanism of the Federal University of Rio de Janeiro in March 2016; and similarly to its fourth edition, entitled *Desenho (...) Cidade (...) Reminiscência* (Drawing [...] City [...] Reminiscence), held at the Superior Technical School of Architecture of Madrid in May 2017 – can be described, based on its objectives:
>
> 1. To discuss how Drawing in/of the City and the elements that identify it (geographical area, inhabitants, natural landscape and/or built landscape; present, desired or memorable facts and data) are represented and identified through the presence and/or action of the body, in the form of gestures, movements, interventions, displacements or permanence.
>
> To recognize the body in/of the city through drawing is an intention.
>
> 'A arquitetura é a única arte cujas obras exigem ser percorridas fisicamente. Só ela exige deslocamentos, percursos, desvios que implicam o envolvimento de todo o corpo e que não podem ser substituídos pela percepção visual isolada.' (Architecture is the only art whose works require to be physically traversed. It alone requires displacements, pathways, detours that imply the involvement of the whole body and cannot be replaced by isolated visual perception. Françoise Choay, *A Alegoria Do Patrimônio*, 2001, p. 258).
>
> 'Eu não sou um "ser vivo" ou mesmo um "homem" ou mesmo "uma consciência", com todos os caracteres que a zoologia, a anatomia social ou a psicologia indutiva reconhecem a esses produtos da natureza ou da história – eu sou fonte absoluta.' (I am not a 'living being' or even a 'man' or even 'a consciousness', with all the characteristics that zoology, social anatomy or inductive psychology recognize in

these products of nature or history – I am the absolute source. Maurice Merleau-Ponty, *Fenomenologia Da Percepção*, 2ª ed., São Paulo, Martins Fontes, 1999, p. 3).

2. To problematise the association between Drawing and City from the starting point of the perception of the Body, assuming this mediation as a condition for the particular construction of that relationship.

'Nossos corpos e movimentos estão em constante interação com o ambiente; o mundo e a individualidade humana se redefinem um ao outro constantemente. A percepção do corpo e a imagem do mundo se tornam uma experiência existencial contínua; não há corpo separado de seu domicílio no espaço, não há espaço desvinculado da imagem inconsciente de nossa identidade pessoal preceptiva.' (Our bodies and movements are constantly interacting with the environment; the world and human individuality redefine each other constantly. The perception of the body and the image of the world become a continuous existential experience; there is no body separated from its abode in space, there is no space unrelated to the unconscious image of our aware personal identity. J. Pallasmaa, *Os Olhos Da Pele*, 2011, p. 38.)

'[...] o sentido intelectual não é o teste final; o teste é a experiência da arquitetura – a arquitetura precisa ser experimentada pelo público (espaço e corpo atuando juntos, aceitando diferentes ângulos, vistas, texturas, cheiros e detalhes).' (The intellectual sense is not the final test; the test is the experience of architecture – architecture needs to be experienced by the public (space and body acting together, accepting different angles, views, textures, scents and details. Steven Holl, *Cuestiones De Perception*, 2011).

3. To identify the presence of the Body in the Representations/Drawings of the City, submitting this event or phenomenon to analysis, aiming for cognitive production.

'[...] Essa era a imagem idealizada do corpo, segundo o seu código, elaborado a partir de diversas fontes e práticas de há muito estabelecidas, como se verá. Baseados nela, seus princípios moldaram o interior do Pantheon, onde o quadrado estava inscrito dentro do círculo, e inspiraram, séculos depois, os desenhos de Leonardo e Serlio. Sua crença fundamentava-se na escala do corpo humano, com base na qual o arquiteto devia modelar o prédio a ser construído. Mais: a geometria humana seria o indício de como uma cidade deveria ser.' (This was the idealized image of the body, according to its code, drawn from various long established sources and practices, as we will see. Based on it, its principles shaped the Pantheon's interior, where the square was inscribed within the circle, and inspired, centuries later, the drawings of Leonardo and Serlio. His belief was based on human body's scale, on the basis of which the architect was to model the building to be constructed. More: human geometry was the clue as to how a city should be. Richard Sennet, *Carne E Pedra: O Corpo E A Cidade Na Civilização Ocidental*, 2010, p. 95).

Thus, the fifth International Seminar *Arquitecturas-Imaginadas: Representação Gráfica Arquitectónica e 'Outras-Imagens'* (Imagined Architectures: Architectural Graphic Representation and Other Images) dedicated to the theme *Desenho (...) Cidade (...) Corpo, Habitando a Terra* (Drawing [...] City [...] Body, Inhabiting the Earth), saw the participation of teachers, researchers and specialists (integrating the abovementioned Research Project, and others similar ones) who, overall, have been dedicating their scientific work to the study of this subject and who, through talks, conferences and round-tables open to public discussion and debate, contribute to the clarification of, reflection on and critique of these topics – from our point of view so evidently paramount to the knowledge transfer within the disciplinary area of Drawing applied to the education of Architecture, Urbanism and Design – the texts published in this book mirror that and underwent a double blind review carried out by an International Scientific Committee.

Lastly, I would like to thank, as Editor of this book, all the participants and the partner institutions.

Pedro António Janeiro

Drawing (...) City (...) Body, Dwelling on Earth – Janeiro (Ed.)
© 2019 Taylor & Francis Group, London, ISBN 978-0-367-02602-8

Committees

CONFERENCE CHAIRS

Pedro António Janeiro
CIAUD, Lisbon School of Architecture, Universidade Lisboa, Lisbon, Portugal

Myrna de Arruda Nascimento
Faculty of Architecture and Urbanism, University of São Paulo, Brazil
University Center Senac

Ricardo Ferreira Lopes
Faculty of Architecture and Urbanism, Federal University of Juiz de Fora, Brazil

ORGANIZING COMMITTEE

Pedro António Janeiro
CIAUD, Lisbon School of Architecture, Universidade Lisboa, Lisbon, Portugal

Fernando Moreira da Silva
CIAUD, Lisbon School of Architecture, Universidade Lisboa, Lisbon, Portugal

Jorge Cruz Pinto
CIAUD, Lisbon School of Architecture, Universidade Lisboa, Lisbon, Portugal

Maria Dulce Loução
CIAUD, Lisbon School of Architecture, Universidade Lisboa, Lisbon, Portugal

Miguel Baptista-Bastos
CIAUD, Lisbon School of Architecture, Universidade Lisboa, Lisbon, Portugal

Ana Moreira da Silva
CIAUD, Lisbon School of Architecture, Universidade Lisboa, Lisbon, Portugal

Myrna de Arruda Nascimento
University of São Paulo, Faculty of Architecture and Urbanism, Maria Antonia University Center, São Paulo, Brazil

José Nicolau Gregorin Filho
University of São Paulo, Faculty of Architecture and Urbanism, Maria Antonia University Center, São Paulo, Brazil

Ricardo Ferreira Lopes
Federal University of Juiz de Fora, Faculty of Architecture and Urbanism, Minas Gerais, Brazil

INTERNATIONAL SCIENTIFIC COMMITTEE

Pedro António Janeiro
CIAUD, Lisbon School of Architecture, Universidade Lisboa, Lisbon, Portugal

Fernando Moreira da Silva
CIAUD, Lisbon School of Architecture, Universidade Lisboa, Lisbon, Portugal

Jorge Cruz Pinto
CIAUD, Lisbon School of Architecture, Universidade Lisboa, Lisbon, Portugal

Maria Dulce Loução
CIAUD, Lisbon School of Architecture, Universidade Lisboa, Lisbon, Portugal
Miguel Baptista-Bastos
CIAUD, Lisbon School of Architecture, Universidade Lisboa, Lisbon, Portugal
Ana Moreira da Silva
CIAUD, Lisbon School of Architecture, Universidade Lisboa, Lisbon, Portugal
Javier Segui De La Riva
Department of Graphic and Architectonic Idea, Escuela Tecnica y Superior de Arquitectura, Politecnico de Madrid, Spain
Javier Mosteiro
Department of Graphic and Architectonic Idea, Escuela Tecnica y Superior de Arquitectura, Politecnico de Madrid, Spain
Javier Raposo
Director of the Department of Graphic and Architectonic Idea, Escuela Tecnica y Superior de Arquitectura, Politecnico de Madrid, Spain
Ana Esteban Maluenda
Department of Graphic and Architectonic Idea, Escuela Tecnica y Superior de Arquitectura, Politecnico de Madrid, Spain
Teresa Fonseca
Faculty of Architecture, University of Porto, Portugal
Ana Godinho Gil
Universidade Nova de Lisboa, Lisbon, Portugal
Carmine Gambardella
President of Research Center Benecon, UNESCO Chair on Landscape, Cultural Heritage and Territorial Governance, Italy
Alessandro Ciambrone
Research Center Benecon, UNESCO Chair on Landscape, Cultural Heritage and Territorial Governance, Italy
Vito Cardone
Presidente della Unione Italiana del Disegno, UID, Italy
Salvatore Barba
Università di Salerno/Unione Italiana del Disegno, UID, Italy
Francesca Fatta
Unione Italiana del Disegno, UID/Facoltà di Architettura, Università Meditaerranea di Reggio Calabria, Italy
Cherubino Gambardella
Catedrático da Seconda Università di Napoli, Italy
Antonello Monaco
Facoltà di Architettura, Università Meditaerranea di Reggio Calabria, Italy
Francesco Moschini
A.A.M., Architettura Arte Moderna, Rome, Italy
Francesco Maggiore
Politecnico di Bari, Italy
Vincenzo D'Alba
Fondazione Gianfranco Dioguardi/Politecnico di Bari, Italy
Walter Angelico
Departamento de Arquitectura, Politecnico de Palermo, Italy
Santo Giunta
Departamento de Arquitectura, Politecnico de Palermo, Italy

Giuseppe De Giovanni
Departamento de Arquitectura, Politecnico de Palermo, Italy
Antonio Conte
Facoltà di Architettura di Matera, Università degli Studi della Basilicata, Italy
Angela Colonna
Facoltà di Architettura di Matera, Università degli Studi della Basilicata, Italy
Antonella Guida
Facoltà di Architettura di Matera, Università degli Studi della Basilicata, Italy
Pedro da Luz Moreira
School of Architecture and Urbanism, Fluminense Federal University, EAU/UFF, Brazil
Marcelo de Andrade Romero
Pró-Reitor de Cultura e Extensão, USP, Brazil
Myrna de Arruda Nascimento
Faculty of Architecture and Urbanism, University of São Paulo, Brazil/University Center Senac
Feres Lourenço Khoury
Faculty of Architecture and Urbanism, University of São Paulo, Brazil
Luís Antônio Jorge
Faculty of Architecture and Urbanism, University of São Paulo, Brazil
Artur Rozestraten
Faculty of Architecture and Urbanism, University of São Paulo, Brazil
Eneida de Almeida
Faculty of Arts, Communication and Education Sciences, University of São Judas Tadeu, Brazil
Fernando Vázquez Ramos
Faculty of Arts, Communication and Education Sciences, University of São Judas Tadeu, Brazil
Ana Aparecida Barbosa Pereira
Faculty of Architecture and Urbanism, Federal University of Juiz de Fora, Brazil
Fernando Tadeu de Araujo Lima
Faculty of Architecture and Urbanism, Federal University of Juiz de Fora, Brazil
Mauro Santoro Campello
Faculty of Architecture and Urbanism, Federal University of Juiz de Fora, Brazil
Ricardo Ferreira Lopes
Faculty of Architecture and Urbanism, Federal University of Juiz de Fora, Brazil
Ralf Castanheira Flores
University Center Senac, Brazil
Valéria Fialho
University Center Senac, Brazil
Andréa da Rosa Sampaio
School of Architecture and Urbanism, Fluminense Federal University, Brazil
Lelia Mendes de Vasconcellos
School of Architecture and Urbanism, Fluminense Federal University, Brazil

Drawing (…) City (…) Body, Dwelling on Earth – Janeiro (Ed.)
© 2019 Taylor & Francis Group, London, ISBN 978-0-367-02602-8

Art and Morphology: The legacy of Camillo Sitte

A. Barbosa
Faculdade de Arquitetura e Urbanismo, Universidade Federal de Juiz de Fora, Juiz de Fora, Brazil

R. Duque
Instituto de Arquitetura e Urbanismo, Universidade de São Paulo, São Carlos, Brazil

ABSTRACT: Camillo Sitte did a meticulous analysis of the urban fabrics of ancient cities. He sought to point out the reasons, from the aesthetic point of view, for people to feel "good" in these cities, highlighting the role of the asymmetry of open public spaces. In this study, we present a critical analysis based on his work "City Planning According to Artistic Principles", in the search for a constant discussion of the author's understanding of the city as an aesthetic object, focusing on what he shows as art in the old city or part of it; we seek to understand the meaning of Aesthetics presented by Sitte in attributing such artistic values to the old city or to a part of it. For this reflection, we support our arguments on authors such as Françoise Choay, Petra Lamers, Gonzáles-Varas, and especially Ivone Salgado; and in authors such as Luigi Pareyson through his work "Il problemi dell'estetica" and Marc Jimenez in "Qu'est-ce que l'esthétique?" in the search for understanding what is Aesthetics and their presence in Sitte's work.

Keywords: squares, old cities, aesthetics, urban design, Sitte

1 A METHOD OF UNDERSTANDING

Camillo Sitte can be considered as an important theorist of nineteenth-century urbanism, as well as Ildefonso Cerdà. According to Callabi (2000), in his analyzes of the urban space Sitte, besides using the static observations according to traditions of the history of art, introduces a new element in these analyzes: time. The author also states that Camillo Sitte's theses published in his book der Städtebau "exerted enormous influence on an entire generation of European urbanists", even to the present day, with a significant rescue after the 1960s. As Choay (2001) says, Sitte does not propose to copy or reproduce the old urban settlements. For this, he "resorts to a rational and systematic treatment of morphological analysis". And in this process the perspective effect of irregular spatial boundaries is explored permanently (Lamers, 2003). The formal reading of Camillo Sitte occurs "from the perspective of the psychology of the perception of the aesthetic effect of the relations of proportions between the monumental buildings and the squares [...]. From phenomenology, Sitte defines a typology of square" (Lamers, 2003), which represents a system of closed squares in ancient times. For him the fundamental question is to define the specific structures of a special landscape, constructed in a three-dimensional way presenting its visual and sensorial qualities. In this sense, we emphasize that for the perception of the closure of an open public space, we can consider geographic questions, as presented by Charre (1990) on the importance of topography in the "artisticity" of the environment

> [...] the role accorded to aesthetics does not limit itself to urban furnishing, nor to the decoration of the facades, but invests the elaboration of the entire topographical organization (Charre, 1990: 221).

We also recall that for Sitte (1992) "the nonconscious and natural artistic sense" that has always organized urban spaces is determined at the same time by the transforming norms of historical cultures and by a stable mental organization. "The invariant, which must allow the formulation of universal principles and laws usable for the elaboration of the constructed, is then in the domain of psychology" (Choay, 1980). In Sitte (1992), there is a preliminary analysis of the spatial dispositions that ancient cities take from their beauty (Figures 1 and 2). He describes and explains, from the ancient city to the Baroque, how different configurations of space have ceased to radiate beauty; the ancient city can give lessons (Choay, 2001). We believe, therefore, that Sitte is also positioned as a philosopher in his reflections and problematizations of urban aesthetics.

This results in the notion of the artistic environment of cities, initially an evocative concept. From that concept and the planning schemes, Sitte (1992) was able to elaborate, in the late nineteenth century, a set of urban intervention techniques, with aesthetic parameters.

For Sitte, the change from the old city to the contemporary city is due to a change of culture, to an irreversible transformation of mentalities (Sitte, 1992). The Viennese author compares plans and effects in perspectives of several cities with his exhaustive method, animated by the will to prove the coincidence of facts and theory, signaling a scientific focus in his studies. He is an urban art theorist and performs morphological analysis (Choay, 1980). From Camillo Sitte's reading of the part to "the whole," it is the Roman Forum which he designates as the principal parameter, and adopts the aesthetic rules of edification; not those of the square (medieval, Renaissance or baroque) or of the old city (Lamers, 2003). In Sitte (1992) the creative quality of urban space is more important than the architectural form, emphasizing its artistic character. The author carried out a research to investigate how the public spaces were constructed architecturally, emphasizing essentially that they were "open", because they had no cover and "arranged for meetings in the open air", but visually closed.

> Hence the possibility of its analysis being made from the artistic principles that refer us to the visual arts, from the principles organized visually by the perspective, even for the period in which it was not possible to represent it with the technical domain that the Renaissance reached, but it didn't prevent that the organization of spaces wasn't based on these principles (Salgado, 2004: 5).

In this sense Salgado calls attention for in Greece the assemblies took place in the Ágora that was outdoors. There was a second square in importance that was the market square, also represented by an open and uncovered space. We recall that in the Roman world, public activities were developed in spaces that could be considered hípteros temples, where even the house had an open patio in which the rooms were added around (Salgado, 2004). According to Sitte (1992) the forum is for the city as what the atrium is for the house. The Viennese author reminds us that the main town squares until their time remain faithful to the tradition of the ancient Forum, where an important part of public life was peculiar to them as well as many of the relations with the monumental buildings that surrounded them.

Figure 1. Roman Forum (from Salgado, 2004).

Figure 2. Greek Forum (from Salgado, 2004).

> There remained the desire to gather in some major points of the city more imposing buildings, but in order to keep a closed set. He comments that, if necessary, porches and arches should be placed in the squares to serve as a complete closure of the mural contour of the square, seeking to preserve a marvelous union of the main walls of it (Salgado, 2004: 8).

Salgado (2004) recalls that Vitruvius recommends three ways of arranging the atrium of houses according to their dimensions, always conditioned by the division of the length from which a measure is taken for width. The author still reminds us that "in the work of Camillo Sitte, the attempt to establish proportional magnitudes for the definition of public spaces refers us to the artistic principles of antiquity architecture" (Salgado, 2004: 9). From this we understand that for the architecture of the Renaissance, the resumption of the classical interpretation of Vitruvius, in view of the concepts of symmetry and proportion, are fundamental, in order to translate the theoretical and doctrinal factors to be adopted. Argan (2003), while understanding the refinement of such concepts in their origin, emphasizes that for the "architectural composition of the isolated building, they [concepts] can easily be extended to the articulation of different buildings in a single context and to the harmony of the parts that conform the unitary context of the city".

We understand that there is a clear need for a proportional relationship between buildings in space. For Salgado (2004) the concepts of symmetry and proportion of the classical architecture and the architecture of the Renaissance can be analyzed from a broader perspective, considering the principles of Rudolf Wittkower, in which all the higher civilizations:

> they believed in an order based on numbers and relations between numbers, seeking to establish a harmony between them, both fantastic and mystical, between the universal and cosmic concepts and the life of man (Salgado, 2004: 9).

Camillo Sitte, through the elaboration of studies on the artistic principles of ancient cities, seems to pursue an understanding of the spatial composition of these cities in order to reveal the different historically created systems of proportions. In this context it should be noted that

> the discovery that the musical consonances are expressible arithmetically grounded the theory of beauty in architecture and from there the various systems of proportion proceeded: the geometric proportion, the arithmetical proportion and the harmonic proportion (Salgado, 2004: 10).

It is apparent that Camillo Sitte works to clarify compositional relations of the old squares as opposed to the modern ones, observing the relations of proportion between height and width from the main buildings that structure these squares. As Salgado (2004) reminds us, until the eighteenth century, there was no doubt that the objective norms of proportion were necessary to the work of art. Already in the nineteenth century, "beauty and proportion were no longer considered universal beings", but something belonging to the realm of psychological phenomena that originate in the artist's mind. With this, "beauty and proportion have come to depend on the creative impulse." Modern psychology defends the hypothesis that the search for order and basic harmony is practically consubstantial with human nature and that "order and proportion in the arts imply giving a conscious and intellectual direction to a subconscious impulse" (Salgado, 2004: 13–14).

2 AESTHETIC ISSUES FOR A SITTIAN LOOK

"Der Städte-Bau nach seinen Künstlerischen Grundsätzen", represents perhaps Camillo Sitte's most important work. First published in 1889, when translated into Portuguese can be interpreted as: A Construção das Cidades Segundo seus seus Princípios Artísticos (Brazilian translation), or Planejamento de Cidades Segundo Princípios Artísticos (Portuguese translation), which in our view represent interpretations with different meanings. We understand that planejamento [planning] is different from construção [constructing], and that the insertion of seus [their] in the title of the Brazilian translation makes it even more different from the Portuguese one, because the term clarifies that the "artistic principles" are those of the city itself.

For Sitte (1992) the spatial composition should be established in such a way that the visual perception of the buildings would cause in the observer the aesthetic effect that would send him to a pleasant and comforting, warm and beautiful sensation. In this way, in the expectation of understanding "Aesthetics" in Camillo Sitte we observe that: for Jimenez (2000), the search for the autonomy of aesthetics as a discipline in itself "attests in fact the existence of a particular domain, linked to sensitivity, which finally obtains an official citizenship right in the same way as the different sciences". And in this context Aesthetics participates in the knowledge and increase of knowledge. But at the same time, this new status translates a will to scientificize the universe of the sensible, in other words, an attempt to rationalize, to theorize and to conceptualize a world of affection, intuition, imagination, passion, rebellious to any way of control or coercion (Jimenez, 2000: 86).

We believe that this understanding can bring us closer to the Sittian readings for the beauty of ancient cities, portrayed through the studies of Italian squares and northern Europe. However, Pareyson (1989: 15) tells us that in fact no indication can come from the name "aesthetics" that arose

> in the seventies, when beauty became an object of confused or judgmental knowledge and in the early eighties, when art was impregnated with feeling and it seemed natural to refer the theory of beauty to a doctrine of sensibility and philosophy of art to a theory of feeling (Pareyson, 1989: 15).

For him, since then the term has been expanding so much to designate the theories of beauty and art, as to understand theories that do not refer to beauty, sensation or art and they do not even link art to beauty. But for the research universe of Camillo Sitte, situated at the end of the 19th century, we believe that the term "Aesthetics" comes closer to its conception of the early nineteenth century, and less than that of the twentieth century, when German philosophers tried to distinguish aesthetic, understood as philosophy of the beautiful, a general theory of art. As well as the current concept of Aesthetics that can be understood as metaphysics "which derives a particular doctrine of systematic principles", or as phenomenology "that interrogates and makes speak the concrete data of experience", or as a methodology of reading and criticize works of art, or as a complex of technical observation and precepts "that may interest artists as well as critics or historians" (Pareyson, 1989). However, we consider it relevant to point out that in Sitte (1992) embryonic precepts of phenomenology, which for Merleau-Ponty can be defined as follows:

> *phenomenology is the study of essences, and all problems, according to it, are summed up in defining essences: the essence of perception, the essence of consciousness, for example. Phenomenology is also a philosophy that replaces the essences in existence, and does not think that one can understand man and the world otherwise than from "factiveness." It is a transcendental philosophy that puts the affirmations of the natural attitude in suspense, but it is also a philosophy for which the world is already there, before reflection, as an inalienable presence, and whose whole effort consists to rediscover this naive contact with the world, in order to finally give it a philosophical status (Merleau-Ponty 1994: 1).*

We believe that Sitte approached this principle considerably, but in any case, regardless of which aesthetic approach is linked to the artistic ideals of Sittians, we perceive that they manifest themselves in the form of philosophical debate, contextualized to the problems of the nineteenth century. And for Aesthetics specifically, for Jimenez (2000) it must be considered that "if the term Aesthetics is adopted, none of the theories of art subsequent to Baumgarten will refer to the content of their theses". For to him Aesthetics has developed independently of what defines it as the science of the beauty, in its origin. For Jimenez (2000), Kant is interested in the beauty in itself and also by the taste, the judgment of appreciation applied to the natural beauty and the art beauty. Hegel, however, renounces any metaphysics of art to the benefit of the philosophy of art, centered on the work and art as an expression of truth. "His conception of aesthetics, linked to the evolution of art from antiquity to the Romantic period, has nothing to do with Baumgarten's or Kant's" (Jimenez, 2000: 115).

Hegel's aesthetics live from the memory of Greek art, set up as an inimitable model, at a time when "the idea that Greece could have reached perfection besets the minds of poets and thinkers, especially Hölderlin, Hegel, Marx, Nietzsche and Freud" (Jimenez, 2000: 115). We emphasize that for the author, in the nineteenth century, "aesthetics is autonomous", does not submit itself to transcendental rules, obeys its own precepts, like art itself. But, on the other hand, this autonomy is relativized, for if aesthetics were totally free from reality, art would become useless purely decorative, recreational.

We can understand that there is death of art, or we can go further and get to the death of public spaces as a denunciation of Camillo Sitte. And in this line, we consider interesting to remember that for Argan (1998) the death of the city represents the death of art. It allows us to reflect on the legacy of Sitte as a critic and manifest against the conceptual transformations linked to the art of building space, urban aesthetics, the work of art. Thus, we think it is more coherent to focus on Pareyson's (1989) clarification that aesthetics is philosophical reflection, and therefore "purely speculative rather than normative – it is aimed at defining concepts and not setting standards". So, we can reflect in the following sense: what Camillo Sitte does is to indicate ways for us to perceive the urban universe in order to build the modern city better, without canceling the creativity, but rather resuming, redeeming its attribution to the architect and not to the engineer as was done in the nineteenth century, with hygienic and circulation demands mainly, as Salgado (2004) explains. Sitte studies the artistic environment of the city against the hygienic-sanitary manual. The author presents a reaction against the urban trends of his time in order to provide artistic bases for "city building" and not "urban planning", according to Brandão (2004), and to distinguish urban design from urban planning. Sitte (1992) "explains for the first time analytically the beauty characteristics of historical cities" (Gonzáles-Varas, 1999).

3 FINAL CONSIDERATIONS

Aesthetics cannot pretend to establish what art or beauty should be, but on the contrary, it has the task of accounting for the meaning, structure, possibility and metaphysical reach of the phenomena that present themselves in aesthetic experience (Pareyson, 1984). It is in this proposition that we believe that Sitte has generated its principles of reading and posture in front of the old city as an artistic object to be adequately understood in the way it has come to the present, to a responsible proposal of transformation or adaptation to modern needs without detracting from its old values, but on the contrary, learn from them.

And it's in view of this effort to reflect on problems of beauty and urban art that Aesthetics present in Camillo Sitte is done as philosophy, where according to Pareyson (1984) "aesthetics is not part of philosophy, but the whole philosophy". And for this Pareyson emphasizes that aesthetics cannot be reduced to criticism, or poetics or technique, it's all this at the same time. It presents itself as a purely philosophical inquiry, but manifests itself as "reflection that is built under the aesthetic experience" so it is not confused with philosophy. For Pareyson (1989) aesthetics is philosophy precisely because it is speculative reflection on the aesthetic experience

> *in which all the experience that comes with beauty and art comes in: the experience of the artist, the reader of the critic of the historian, of the technician of art, of the one who enjoys any beauty (Pareyson, 1989: 18).*

We understand that Aesthetics is philosophy, so it is reflection on experience. It has a speculative and concrete character at the same time, as Camillo Sitte presents us in his studies for the construction of cities according to their artistic principles.

REFERENCES

Argan, G.C. 2003. História da Arte Italiana. Tradução de Vilma de Katinshy., São Paulo: Cosacnaify.
———. 1998. História da arte como história da cidade. 4.ed. Tradução de Píer Luigi Cabra. São Paulo: Martins Fontes. (Coleção a).
Brandão 2004. Articulações e desarticulações da historiografia recente da arquitetura e do urbanismo. In: PINHEIRO, E.P.; GOMES, M.A. de F. (org.). A cidade como história: os arquitetos e a historiografia da cidade e do urbanismo. Salvador: EDUFBA. 223 p.
Calabi, D. 2000. Storia dell'urbanistica europea: questioni, strumenti, casi esemplari. Veneza: Paraiva Scriptorium.
Charre, A. 1990. La riorganizzazione estetica e topogefica della città di Lione. In: ZUCCONI, G. Camillo Sitte e i sui interpreti. Milano: Urbanística Francoangeli. p. 221–228.
Choay, F. 2001. A Alegoria do patrimônio. Tradução de Luciano Vieira Machado. São Paulo: Estação Liberdade; UNESP.
———. 1980. A Regra e o modelo. São Paulo: Perspectiva.
Costa, L.A.M. 2004. De Immanuel Kant a Camillo Sitte as noções de história, estética e projeto em: a construção das cidades segundo seus princípios artísticos. In: CONGRESSO INTERNACIONAL DE HISTÓRIA URBANA, 2004, Agudos. Camillo Sitte e a circulação das idéias de estética urbana. Europa e América Latina: 1880–1930. Agudos/SP: Cultura Acadêmica. 1CD-ROM.
Gonzáles-Varas, I. 1999. Conservación de bienes culturales: teoria, historia, principios y normas. Madrid: Catedra.
Hegel, G. W. F. 1999. Cursos de Estética I. Tradução de Marco Aurélio Werle. São Paulo: Edusp (Clássico).
Jimenez, M. 2000. O que é estética? Rio Grande do Sul: Editora Unisinos (Coleção Focus).
Lamers, P. (Coord.). 2003. Teoria da arquitetura até aos nossos dias: 117 tratados apresentados em 89 estudos. Tradução de Maria do Rosário Paiva Baléo. Lisboa: Vernáculo.
Merleau-Ponty, M. 1994. Fenomenologia da Percepção. Tradução Carlos A. de Moura São Paulo: Martins Fontes. (Coleção Tópicos). 662 p.
Pareyson, L. 1989. Os problemas da estética. São Paulo: Martins Fontes. (Coleção Ensino Superior). 180 p.
Salgado, I. 2004. Uma leitura sobre a concepção estética de Camillo Sitte: a construção do espaço perspéctico. In: CONGRESSO INTERNACIONAL DE HISTÓRIA URBANA, 2004, Agudos. Camillo Sitte e a circulação das idéias de estética urbana. Europa e América Latina: 1880–1930. Agudos/SPl: Cultura Acadêmica. 1CD-ROM.
Sitte, C.; ANDRADE, C.R.M.A. (Org.). 1992. A Construção das cidades segundo seus princípios artísticos. Tradução de Ricardo Ferreira Henrique. São Paulo: Ática.

Drawing (...) City (...) Body, Dwelling on Earth – Janeiro (Ed.)
 ISBN 978-0-367-02602-8

The drawing of the winding line

A. Godinho
Instituto de Filosofia da Nova, Faculdade de Ciências Sociais e Humanas, Universidade Nova de Lisboa, Lisboa, Portugal

ABSTRACT: This text aims at analysing the conditions that make the drawing be seen as a design of creation. And, as a potential field, a way of making accurately everything visible, we have to ask again: what is drawing? Why is drawing so different from other arts? Although we have a community of arts which share some common problems, we will just highlight one of them. According to Deleuze and with a "logic of sensation" drawing starts to be seen in a lineage that goes from Leonard de Vinci without forgetting Ravaisson, Bergson, Valéry, through the development of a winding and frenetic line (Wörringer's line) until its inevitable relationship with the body and with life. Drawing is similar to a seismograph in catching both the movements and the most secret rhythms of the universe. When reading "The Invisible Cities" by Italo Calvino we will draw other lines. What such a mystery does drawing have that appears to become an obsession and let Pliny the Elder cite Apelles: "nulla dies sine linea"; or Artaud state that the secret is saying nothing and representing absolutely nothing, and Van Gogh ask "What is drawing? How does one get there? It's working one's way through an invisible iron wall that seems to stand between what one feels and what one can do. How can one get through that wall?"

Keywords: Drawing, line, trace, body, magic

1 THE PRIMORDIAL LINE OF THE UNIVERSE

This text is a sketch, a preliminary drawing, an excuse that enables us to ask once more the same question we have asked countless times: what is thinking? And that goes to another question: what is drawing? What is a drawing? Why is drawing so different from other arts? Why does it have that quality which makes it possible to reveal what, according to Hesiod, Gods will have hidden and that makes men live?

We think that Valéry is right when he questions himself about drawing and asks if it is not the major obsession of the spirit, an obsession that is shown through the movement of the hand that draws or makes a sketch. Creating is always producing lines. "Drawing is free", a free design of the spirit, an unconscious mental operation and at the same time design of the eye, which more than see, becomes mediumistic design of the eye, which beyond time and space captures that hidden thing and draws it (Duchamp).

Leon Battista Alberti (1406–1472), to be brief, privileges a conception of drawing in which the contour borders an object making it visible. Giorgio Vasari (1511–1574) comes with the idea of shape, an inheritance from Plato and Aristotle. *Disegno,* is something intelligible and sensitive. This concept expresses how the matter is made. The drawing is the place where we create any forms according to established codes. It is also a place where we draw and sketch our thoughts and imagination in a movement of previous exploration and preparation (for other forms of art, for example). The drawing will find the line and the movement of its artist, through which he can express himself.

But it is Leonard de Vinci (1452–1519) who brings us new ideas and essential differences, which will change the relationship between body and matter in the drawing: movement, rhythm, incompleteness, freedom to express what we want, winding (serpentine) lines that carry potentialities, experimentations, lack of any contour. The drawing becomes autonomous and *sfumato*, fluid, zone of vibration. The forces and lines that go through it have no substance or matter; carried by the drawing, they go through the space and meet together between the shadow and the light. The drawing, judgement of the eye, without the interference of the mind, is like a mirror that reflects every single object that is before it without knowing them (De Vinci, 2004: 273). For Leonard, the secret of the gift of drawing lies in the discovery, in each object, in every tiny detail, of a certain flexuous line, which resembles a generator, and which is not one of the visible lines of the figure. A line that is less perceptive by the eye than thought by the spirit. Drawing is an obsession as it makes appear in each one of us the "individual meander" that in Bergson's view makes our individuality alive. Obsession that searches behind the lines for which is not representational.

We always draw to give life or to release it from down there where it is imprisoned, to draw lines of flight. Deleuze will call it the line of "continuous variation" like the Wörringer's gothic line or "wizard line", without contour, always unfinished. We can also call it primordial line, of the universe that goes

> "to reach that point where the associative chain breaks, leaps over the constituted individual, is transferred to the birth of an individuating world. And it is no longer a matter of saying: to create is to think – but rather, to think is to create and primarily to create the act of thinking within thought" (Deleuze, 2000: 111).

2 THE POWER OF THE MAGIC

The drawing starts in the first line of its own beginning. We would like to notice the way this line is conceived, and its relationship with the other ones, its implications in painting, sculpture, architecture, in the body, in life, in the city.

Italo Calvino in the beginning of "*Invisible cities*" (Calvino, 1974) tells us that we cannot be sure about everything Marco Polo says when he describes the cities he has visited (the way he has described them to Kublai Kan and has drawn them to us). In life, "there is a moment" in which the "sense of emptiness" reaches us, a moment of "dizziness that makes rivers and mountains tremble" that captures us and "flakes the wax of seals". It is in that moment that Marco Polo can draw the cities, a moment that comes through what we have never heard or seen or found out. By that time we ask the same way Van Gogh did: "what is drawing? How does one get there?" And he answers his question saying that drawing is how we open a "way through an invisible iron wall" (Van Gogh, 1996: 451) (letter quoted by Artaud). Only through Marco Polo's reports ("accounts"), according to Calvino, *"was Kublai Kan able to discern, through the Walls and towers destined to crumble, the tracery of a pattern so subtle [filigree] that could escape from the termites' gnawing" (Calvino, 1974: 6).* But nothing can assure us that we can ever escape from that. We have found in the tracery of a pattern so subtle drawn by Marco Polo, who we see now not as a merchant but as a draughtsman, the possibility of going through that invisible iron wall. This is one of the designs of the drawing.

The filigree is the line or the tracery of a pattern so subtle we follow when we are "meditating, lost in thought", line that unfolds and improvises at its own risk. An imaginary line that gets lost in "unfamiliar quarters of distant cities", and the more it gets lost, the more it changes and will change while moving forward. A draughtsman like Marco Polo goes into a strange city and "sees someone in a square living a life or an instant that could be his". The drawing becomes the place where one could be and that is known to widen endlessly. Drawing to find out the future. After we draw the first tracery of a pattern so subtle, it becomes another and others. The filigree is able to rebuild itself, to replace itself, to move away, to go backwards; endless movements are possible. But, just like the first city, this line has neither "name nor place" or a connecting thread. It is a line of nothing, dreamt in an uncertain and winding north that needs to be decoded and seen. Lines like cities seem to be "work of the mind or of chance, but neither the one nor the other suffices to hold up their walls" (Calvino, 1974:44), as words are not enough, what are we supposed to believe in then?

In *50 drawings for murder magic* Artaud will say:

> *"... These drawings, /are not/the representation/or figuration/of an object/of a state/ of mind or heart,/of an element/or an event/of a psychological kind, /they are purely/ and simply/the reproduction on/the paper/of a magical action/that I have performed/in true space/with the breath of my/lungs/with my hands/with my head/and my two feet/ with my torso and my/arteries, etc." (Artaud, 2008, 16)*

The drawing has therefore this virtue: we can "walk" in it with the whole body and "soul" that have created it with magic the same way we can walk through the cities "in thought", running the risk of getting lost, "stop and enjoy the cool air, or run off." The line is thus a "sensation line", something that you feel (*affectus*), not a sensational line, an obstinate, fierce, vital line, pure immanence, or "pure power", potential as *puissance* (the power to act and create, the sensibility of force), "a life" which expresses through minor events, a whole series of slight differences, reproductions of a "magical action", which represent nothing. The line doesn't separate from the body and feels the same vibrations that go through it – the body-bundle of forces becomes line-bundle of forces, which don't have either a subjective or a representative nature.

And when it becomes too visible and is unable to go forward, when it is no more fluid or *sfumata*, it may be drawn again (as Marco Polo does) in order to see it differently. Coming back to the hands (each one of them corresponds to "a shift of mood, in their alternation and repetition"), with them we draw once more the "monuments, markets, costumes, fauna and flora" and everything else words are unable to catch.

If the line, the tracery of a pattern so subtle is like Marco Polo's city thus "the city exists and it has a simple secret: it knows only departures, not returns", and if it doesn't exist, it has probably never existed. "Your cities do not exist. Perhaps they have never existed. It is sure they never exist again."

But the draughtsman sharpens his look: "If you want to know how much darkness there is around you, you must sharpen your eyes, peering at the faint lights in the distance." Up there, where the look goes, the strength of the line appears, the grey point, and up there it is drawn in accordance with "a perfect pattern", and Calvino goes on: "Amid the surge of the elements, a splendid hard diamond takes shape, an immense, faceted, transparent mountain." The line appears, in an unstoppable process the other supposed lines cannot keep. The draughtsman, Marco Polo, draws, obsessed, with a different line, "made of exceptions, exclusions, incongruities, contradictions", and he concludes that "if such a city is the most improbable", then we need, we believe, to increase the probability of that line existing. But how far can the draughtsman go? He may not go beyond a certain limit, otherwise, e he "would achieve cities too probable to be real", and if he does? If he goes beyond a certain limit, what does he get?

Many difficulties still remain, but even without any guarantees:

> *"Marco Polo describes a bridge, stone by stone.*
> *– But which is the stone that supports the bridge? – Kublai Khan asks.*
> *– The bridge is not supported by one stone or another,*
> *– Marco answers, – but by the line of the arch that they form.*
>
> *Kublai Khan remains silent, reflecting. Then he adds:*
> *– Why do you speak to me of the stones? It is only the arch that matters to me.*
> *Polo answers: – Without stones there is no arch." (Calvino, 1974: 82)*

Polo and Calvino draw in filigree not to lose their "*Invisible cities*". In order to be able to draw the world with words and actions, they need that everything they see and do makes sense "in a mental space where the same calm reigns as here, the same penumbra, the same silence streaked by rustling of leaves" (Calvino, 1974: 103). "As here" is the garden that only exists "in the shadow of our lowered eyelids". Whenever we squint our eyes, we can see more clearly. Through the line of half-open eyelids we separate, differentiate, trace, draw, that is, by losing the usual references, "we cannot know which is inside and which outside" the "hanging garden" or "wasteland". But when "we draw some conclusions" it's impossible to know if the thing turns to the mind or outside. The invisible line that rules the drawing seems to be always about to be discovered and to become whirling, slow, frantic, proliferating, in a spiral shape or about to implode, *mise en abîme.* But, seeing an object is not the same as drawing an object. Kublai Kan just wants to see and that is why he drives his obsession over the edge, to the definitive conquest: the "nothingness…"

Perhaps Marco Polo reaches an "oasis hidden in a fold of the desert" if that is his city. He may find a way of expressing by gestures what he has drawn in his spirit, the multiplicities to come, since he draws as Calvino (who was an excellent draughtsman when he was young) the text that is born continuously through "signals one sends out, not knowing who receives them". He draws in the folds, in the bends, where the "infinite line of virtual inflection" exists. Between matter and soul, between heaven and earth, openings for the slightest difference do not stop to come up not even by "an opening in the midst of an incongruous landscape, a glint of lights in the fog", by hands painted at the bottom of a cave. There, in "instants separated by intervals", in filigree, *"the city toward which my journey tends is discontinuous in space and time, now scattered, now more condensed, you must not believe the search for it can stop" (Calvino, 1974: 164)*. What are Marco Polo and Calvino looking for when the first one draws with gestures in the air and the latter with words?

Marco Polo's drawing, as every drawing, will not stop finding that intrinsic singularity, which doesn't resend to any coordinate, and which is the beginning of the world. Paul Klee "will call it the place of cosmogenesis", the "non-dimensional" point, "between dimensions". He will also state that, in early times, when writing and drawing coincide, the line was the first element. And Marco Polo will also say "perhaps, while we speak, it is rising, scattered, within the confines of your empire". Klee's spontaneous and dynamic line, Calvino's filigree, Artaud's magic action, Wörringer's non-organic powerful life are the real atom, the elastic point, which Deleuze will call line of infinite inflexion or celestial line of flight. Wörringer, Deleuze says, thought he had found his formula for this frenetic line: "it is a life, but the most bizarre and intense kind of life, a *nonorganic* vitality" (Deleuze, 2003: 129).

The way Italo Calvino reads the world goes a thousand times through what is not written and not said, which the drawing and the visual images or the music are more capable of opening, that is to say, to express. Listening to the lines, going from the ear to the eye, passing by the hand. And after the hand, what can be found? Because it is the own universe of the drawing that is drawn, that is redefined at any moment, that makes its own experiments from countless ways, with myriad styles and a prodigious imagination that go through times. And a last question in the last but one penultimate page of the "*Invisible Cities*" "– You, who go about exploring and who see signs, can tell me toward which of these futures the favoring winds are driving us?" (Calvino, 1974: 164).

In his last fable written in 1980 Calvino tells us that in a beautiful day the Line was very angry, as she had to obey to the rules of the Hand. Therefore, she uses her trickery to take control over the situation. She tells the Hand she wants to help, as a sign of acknowledgment, by drawing a hand. The Hand let herself guide by the Line, becoming submissive. The Line rejoices as she is certain she is the only one that can explain to the Hand what a hand really is. But the hand that was drawn, will not cease to draw some hands in different positions, since it was unable to use up all the existential possibilities of the drawing hand (Escher "Drawing Hands", 1948). At the end, the Line believes she is the only one to lead the game, without realizing she is more governed and better controlled than before. The Hand, on the contrary, is convinced she has found out her own essence ignoring that in the absence of the Line she wouldn't really know that she exists.

The drawing, on the contrary, seems to have that quality that makes it live since its origin of "invisible reasons", which make it free. The same reasons Kublai Kan has heard Marco Polo saying: "The invisible reasons which make cities live, through which perhaps, once dead, they will come to life again." (Calvino, 1978: 136).

3 THE INVISIBLE REASONS

The invisible reasons we seek and suspect of being the main elements of the drawing do not stop showing up between heaven and earth, matter and soul, body and spirit. On the walls of the prehistoric caves we find extraordinary traces of that puissance, something able to catch the movements and the most secret and the finest rhythms of the universe.

Jean Clottes and David Lewis-Williams, in the nineties (1996), put forward an idea that still remains nowadays in spite of having aroused considerable controversy. An idea that brings the prehistoric parietal art closer to shamanism, (this theory had already been suggested by two experts: Kirchner and Lommel), and that we formulate our own way by asking: what if those reasons that we miss were in another world? In the world that the shamans (through trance) drew on the walls of the caves and thought to be an underground way of appropriation and communication of forces, control and incorporation that are absolutely essential to an introduction and meeting with the world.

Or, following Pascal Raux's point of view, who in 2004 resumes the idea of the wall as a door, a passage or a porous frontier between the world of the living and the world of the spirits or of the supernatural. The way Pascal Raux "reads" the drawings sends us to the origins of art. From animist beliefs, the author reinforces the idea of animism as being the first religion of humanity, which consists of believing that everything and every being has a living soul. The soul is something different from the body it takes up and which it can leave temporarily during sleep or trance. The first beliefs thought the dream was a phenomenon where we can see what has vanished from our life. The dream was another world that was closely linked to life just like the spirit that leaves the bodies when we die is superior to the body and outlives it. Ancestor worship has the same belief. If drawings are related to rituals of trance, the shaman is the one who draws and can travel to and in another world, when he is in an altered state of consciousness. We cannot help agreeing with Valéry "there is a huge difference between seeing one thing without a pencil and seeing it when it is being drawn" (Valéry, 1938: 63).

Drawing allows subject and object thus to change their status and their nature. We draw the first traces, curves and lines, contours of animals, negative and positive hands, *in sfumato*, figures, silhouettes and uncompleted bodies. We apply pigments using some techniques of a great complexity, we project blows and fluids from the bodies. They are the main traces that come from a will, an intention that is magical or animistic, divinatory or that adjures us.

Drawing for the first time is setting a rhythm, inscribing the movement in acts and events, in traces that are addressed to beings from other worlds, other times, and that define ways through, possibilities of a free and expressive gesture. In this first communication with the world or with the worlds drawing was actually there.

In the crucial moment of the genesis a will of saying and showing our own way of being (style) in the world was felt. "A will of art" or a "will of the shape", as Maldiney notices quite well in "*Regard Parole Espace*". And we still add another will: a will of the forces that fixes materially a time and a space of the past, of the memory and the imagination, to create the present and the future. Going beyond a certain limit is overflying that sinuous line of the primordial inflection of the drawing, which from the beginning seems to be linked to magic. We are not exaggerating when it is said that drawing is magic. Jean Clottes will even state that magic only worked with drawing: the hands drawn on the rock went through it to travel as far as the world of the spirits (as we can see in the Chauvet cave, discovered in 1994, or in Gargas in 1906, or even more recently in Borneo, 2006).

A lot of studies about the eastern handwriting lead to the same conclusion. This handwriting might also make the passage and the communication to the world of the spirits, of the Gods and of the ancestors. Through the drawn surface, through the rituals and the chants, which accompany it, death has been abjured. Perhaps through a "magical action", as we have seen with Artaud and that Fernando Pessoa also describes in the same way:

> *"This action through which something is created is called a magical action, and we call magic the art through which that action is possible/ is achieved. And as the «non nisi parendo vincitur» can be applied here in this context, we need to know how we can control the external force in order to transform it, so that magical action can exist" (Pessoa, 1968: 44).*

The parietal art as well as the eastern handwriting (we could not develop) are some of the fundamental evidence from this beginning of the sinuous travelling line, the one which has never lost its purpose and that is the most unpredictably necessary. If drawing is giving life and venture in it, then "without stones there is no arch". We can speculate and not knowing much more than our own interpretation and theories can reach, but drawing, as well as sensation, is a certainty that confirms the world's reality. And we can say about its origin fair words that Maldiney cites: "either it is here the place where light was born or it is here the place where it reigns freely, after having been taken captive" (Maldiney, 1999: 163).

Far from intending to depict or present a finished product (a figuration, an image close in itself), the drawing is an art of passage, of movement between forces and shapes, matters and signals. Therefore, drawing is said to be a preparation for painting or for sculpture or architecture; and if it is worthy, it is as a movement outside, between what doesn't have any shape and the shape. That is why in every situation a singular line flows, without any dimension, between the night and the light, *sfumata*.

4 THE INVISIBLE ORIGIN OF DRAWING

The invisible origin of drawing, what the whirling line draws or the difficulty in establishing the distinction between painting and drawing together with the empowerment of the last one regarding writing are some of the worries that made us believe (according to Deleuze) that drawing and arts in general share a common problem. That is why we have looked for an axis, a nexus, in what concerns the possibilities of drawing. In a certain way, we know that drawing is different from painting or writing, but the same physical gesture traces, inscribes and embodies itself in the space and in the spirit.

The drawing is generated and kept when we create a trace, which is not ruled by the visible, but that always begins by the invisible. Really contagious, it becomes an explorer of an incalculable extension, which is still unknown for us, and it does "body to body" with it. Drawing is, as we have seen, a way of opening a path through multiple spaces of heterogeneous surfaces: a "wall of invisible iron" or a wall of a cave. It is the availability of opening to live in a place, an area of imperceptibleness that is always about to begin and about to disappear (a huge number of possibilities are possible either in painting, in architecture, or in the technique), or it is still for Duchamp a means of "surpassing the animal state", "a way out to regions governed by neither time nor space". And, just like a line "never exists alone" (in Matisse's viewpoint) the same happens with the drawing. Autonomous and free, it has the power to keep the virginity of the "first" forces, while it makes its own drawing and goes through the time. Deleuze will say in a fair and reasonable way: "There is a pictorial possibility that has nothing to do with physical possibility and that endows the most acrobatic postures with the sense of balance" (Deleuze, 1994: 164). We draw even before starting drawing anywhere, with any element, either matter or material, as we deeply need or wish to do it.

The drawing waves in a direction that is open as we go further, but in some way it remains inaccessible and problematic. It escapes from any reflection as it avoids the mirror, its own image, the sensational and the spectacular. Kora's lover (Dibutade's daughter) has also escaped from it and has kept himself the way he was. There is a trace that lights what is needed to be seen, just like Marco Polo's drawing that always appears in the place of the missing city and about which he never talks but that is always there. Marco Polo's "visions", which make the mountains tremble and escape from the gnawing of the termites are kept in filigree. Calvino has found a way of expressing for ever by gesture what he drew in his spirit.

There is a trace or a line that only art is able to maintain and that western handwriting seems to have lost, but not the eastern writing/handwriting, or the drawing (we can honestly say the same to music, literature and poetry, which we are not to talk about), as they had not let escape the rhythm, the concrete movement of the cosmos, the expressive freedom, the wish of making the first virgin forces visible. They can keep the power of creation through the hand and the trace, and the possibility the magic action calls and puts into practice. It is possible that through Kora's hand, which came from the night, love has invented drawing. The visions suspended in time and space (what Giacometti called style) travel beyond the limits and are drawn countless times before, after and right now with the power they have to decide and create. And, tirelessly, other hands will come from the bottom of the caves up to the surface and will mend it incessantly, as Penelope did. They are hands that follow the spirit and are closely linked to the movements of the soul, which are movements of the world.

As groups of delicate joints, nerves and ligaments that expand and contract, thumb and index finger mould, move roughly and join together in a unique tension exploring beyond the fingertip, memorising visions, dreams and sliding through rhythms and blows. The drawing "puts its hand on the thought", because it makes it come out from the common paths. The draughtsman "sees" behind the lines we see the line of the unseen movement. For Ravaisson, the drawing begins with the "good judgement of the eye", which definitely "sees", when it lets itself go. And, in this act of abandon or "acceptance", it doesn't let the spirit escape from its fundamental design. This is what the draughtsman should do when he creates. In order to truly see life that expresses itself to the eye in the thinnest and most invisible curves the draughtsman becomes "gentle" and, in an act of "acceptance", he pounces and gets into "the privacy of the work of the nature" (Bergson, 1959: 1472). There he catches the gentle movements through which, according to Ravaisson, we feel, we get a feeling and guess in a kind of "distension" of the spirit". Inspired by Leonard de Vinci, to whom beauty was the established grace, Ravaisson considers, according to Bergson, that "grace" is the most concrete "vision" of things, the major freedom, it is acceptance and love, it is life itself.

Between inside and outside, past and future, shadow and light, in the unique edge lives a line that generates life, whirling "cosmic axis" *sfumato* hidden by the Gods, (that was what Hesiod thought), which waved in the wind in a rhythmic and infinite "individual meandering" like a bamboo or a grass or a *disegno*. The "grace" of the drawing lies in its mysterious and magic quality of making my soul and my body be the same, be just one. It is the drawing as *cosa mentale*, body and mind that really matters: "... and it is the only thing in the world that is preserved. (...) The air still has the turbulence, the gust of wind, and the light that it had that day last year, and it no longer depends on whoever was breathing it that morning." (Deleuze, 1994: 163).

REFERENCES

Artaud, Antonin. 2008. 50 drawings to murder magic. London; New York: Seagull Books.

Bergson, Henri. 1959. Oeuvres. Paris: Ed. du Centenaire/Presses Universitaires de France.

Calvino, Italo. 1974. Invisible cities. New York: Harcourt Brace Jovanovich.

Calvino, Italo. 1994. Romanzi e racconti. Milano: eds Mondadori.

Clottes, Jean & Lewis-Williams, David. 1996. Les chamanes de la Préhistoire. Transe et magie dans les grottes ornées. Paris: Seuil.

De Vinci, Leonard. 2004. Les Carnets de Léonard de Vinci. 2 vols. Paris: Gallimard.

Deleuze, Gilles & Guattari, Felix. 1994. What is Philosophy?. Trans. by Hugh Tomlinson & Graham Burchell. New York: Columbia University Press.

Deleuze, Gilles. 2000. Proust and Signs. The Complete Text. Minnesota: University of Minnesota Press.

Deleuze, Gilles. 2003. Francis Bacon: the logic of sensation. London: Continuum.

Duchamp, Marcel. 1975. Duchamp du signe. Paris: Éditions Flammarion.

Gogh, Vincent van. 1996. The Letters of Vincent van Gogh. Trans. Arnold Pomerans. Ed. Ronald de Leeuw. London: Penguin.

Maldiney, Henri. 1999. Regard Parole Espace. Lausanne: Ed. L'Age d'Homme.

Pessoa, Fernando. 1968. Textos Filosóficos. Lisboa: Ática.

Raux, Pascal. 2004. Animisme et arts premiers. Historique et nouvelle lecture de l'art préhistorique, Fontaine:Thot.

Valéry, Paul. 1938. Degas Danse Dessin. Paris: Gallimard.

Drawing (...) City (...) Body, Dwelling on Earth – Janeiro (Ed.)
© 2019 Taylor & Francis Group, London, ISBN 978-0-367-02602-8

Former "Portuguese" Sub-Saharan colonial Africa: A drawing narrative

A. Milheiro
Department of Architecture History and Theory, Lisbon School of Architecture, Universidade Lisboa, Lisbon, Portugal

ABSTRACT: After the creation of the Colonial Urbanization Office in Lisbon in 1944 by the then minister of the colonies, Marcelo Caetano (1906–1980), a production of urban, architectural and infrastructure drawings for the Portuguese colonial territories started. Many of which had never been applied on the ground. But its existence was fundamental in the colonization of the Estado Novo regime (1933–1974) to reaffirm internally and externally the continuation of the Portuguese political and administrative presence in these colonial territories. From the collection of architectural projects to the custody of the Overseas Historical Archive, in Lisbon, an imaginary journey through an African and Asian landscape is designed (and reinvented) from the heart of the metropolis, Lisbon, in this final period of Portuguese colonization (1944–1974). It is also a visual narrative that reinforces an architectural learning that has different performances until the April revolution, when the doors to the independence, especially the African ones, were finally opened.

The research supporting this paper was initiated in the multidisciplinary research project entitled "Colonial Urbanization Offices: Architectural Culture and Practice" funded by the Foundation for Science and Technology (Fundação para a Ciência e Tecnologia, 2010–2013). [Ref. PTDC/AURAQI/104964/2008] and coordinated by me. The research is currently underway in the project "Coast to Coast – Late Portuguese Infrastructural Development in Continental Africa (Angola and Mozambique): Critical and Historical Analysis and Postcolonial Assessment", also funded by FCT (started in 2016) [Ref. PTDC/ATP-AQI/0742/2014].

Keywords: African architecture, Portuguese Colonial Architecture, Urban plans, Architecture Drawings, Colonial Infrastructures

1 INTRODUCTION

Here the story is told from the colonizer's perspective. This is a calculated retreat from conventional postcolonial studies. This makes city, architectural and infrastructure drawings, materials with a strong ideological potential. It is known, however, that this connotation was not in the technical gaze of the architects and engineers who produced these visions; but it would most likely be the decision of the political power that promoted and approved them.

Is well known that the last phase of Portuguese colonialism in Africa accelerated the occupation processes of the colonial territories. The Estado Novo regime was then responsible for the consolidation of the infrastructure processes initiated in the end of the 19th century and for the implementation of public facilities which transformed the reality of the majority of the overseas cities, with contemporary implications. The contribution to this built landscape came from the Lisbon-based agencies, under tutelage of the Overseas Ministry (such as the very special case of the Colonial Urbanization Office, created in 1944 and with multiple ramifications until the April Revolution of 1974), of the public works local departments (with the progressively installation of specialized technicians since the legislative reform of 1915) and also of private promoters who invested in the construction of facilities (namely leisure and services buildings), hiring professionals from the metropole or already established in Africa.

In the case of the public promotion works, this dynamic of spatial appropriation indicated a strategy of uniformity of the African landscape, which can be interpreted as a way to subjugate the colonized territories. In this paper, a special attention is given to the first way of action: that was, to the work directly dependent on political decisions of the central and provincial governments, built within the scope of the different departments in charge of Public Works in the Portuguese colonial space.

2 PORTUGUESE COLONIAL PUBLIC WORKS AGENCIES

To outline a common itinerary for the architectural culture officially promoted during the Estado Novo dictatorial regime means to describe unequal scenarios. Provinces with a larger dimension, with better economic performance and a greater capability of attraction of "European" immigrants, such as Angola and Mozambique, were distant, in terms of public investment, from the peripheral condition of Cape Verde, Guinea-Bissau and São Tomé and Príncipe.

A systematic reading of the four Development Plans within the scope of Public Works – four plans that, from 1953 onwards, created the financial conditions for the implementation of the main colonial policies – also helped to clarify the options of that investment, focused on territorial infrastructure programs connected to the maximization of natural resources exploration and assisted colonization, potentially European, in Angola and Mozambique.

The colonial war, which spread throughout the three continental provinces (Angola, Guinea-Bissau and Mozambique) in the beginning of the 1960s, contributed to intensify the differences. The specific situations of each region were thus patent in the dominant architectural principles, breaking the homogeneity inscribed in the official strategies. This fact allows us to acknowledge, as a starting point, two distinct evolution frameworks, whose complexities will thicken: Angola and Mozambique were experimenting a more progressive investment, based on industry development and rowmaterials exploitation. The other three regions (Cape Verde, Guinea-Bissau and São Tomé and Príncipe) based on a more subsistence economy.

However, despite the diversity of conditions experienced in each of the African territories, the colonial Portuguese State seek homogeneity. Among the different questions that we have been raising during the research process, we are interested in one: What territory occupation processes are the basis of the African landscape homogeneity in the last phase of the Portuguese colonization?

The answer to this question was directly related to the Colonial Urbanization Office, already quoted, a department/agency of the Colonies Ministry (later Overseas Ministry), created in the aftermath of the Allied victory in World War II, and one of the tools of resistance to the international pressures towards decolonization, due to the visibility of the actions of its technicians. That visibility was often a consequence of a production focused on the creation of images associated with a "developmental" ideal, and not so much the result of the amount of work built or plans implemented.

3 DRAWING STRATEGIES

The different designations of this Public Works agency until its extinction in 1974, were expressed in the changes in the drawing strategies that this was promoting for the city and its buildings. As a result of the multidisciplinary teams of architects, engineers, experts in tropical medicine and climatology who worked here, a colonial, functional and tectonically solid "representation" architecture was developed that evolved in three stylistic phases:

i. First, drawing on the Portuguese popular architecture of the Alentejo and, generally, the southern part of the country, in a "more Mediterranean" version or a picturesque approach, equivalent to the GUC period (1944–1951).
ii. Later, testing more monumental and historicist typologies, associated with dictatorial regimes, reflecting legislative changes that repressed the use of the term "colonial", replaced the original name of this organism by Overseas Urbanization Office (GUU, 1952–1957).

iii. Later on, they allowed themselves to be contaminated by local constructive traditions and rehearsed a first expression of "African nativism", anticipating visions of autonomy and independence. This time, very late and, compared to previous phases, with less concretions on the ground, corresponded to the performance of the Directorate of Urban Planning and Housing of the Directorate General of Public Works and Communications of the Ministry of Overseas (DSUH/DGOP-MU, 1958–1974).

These ideas were also reflected in the town plans, in the proposals for low density – which was thought to be most conducive to the tropics –, then in the ripening of large and monumental avenues, and finally in the integration of the musseques, sanzalas, or "caniços" – spontaneous settlements or the "black" city (though integrating poor European settlers) into the formal city.

4 FUNTIONAL STRATEGIES

Launched by Marcelo Caetano in the end of 1944, the objectives and powers of the Colonial Urbanization Office (Gabinete de Urbanização Colonial – GUC), headquartered in Lisbon, were described in the Decree no. 34:173, of the Colonies Ministry.

The decree defined three areas of action which stand out precisely because of their network mode of operation, enhancing an infrastructural kind of occupation:

i. The Medical-sanitary assistance was the most comprehensive program of this first stage, due to its extension to all the layers of the colonial population. It encompassed central and regional hospitals, for the European colonists, and primarily located in urban areas, as well as facilities for the African rural communities, whose sanitary intervention plans included the installation of standardized services (sanitary posts, rudimentary infirmaries or maternities).
ii. Houses for employees, an essential program in the settlement of European colonists and mostly aimed at the "white" population. The single-family residences were directly related with the choice of the Garden City model, which informed the majority of the urbanistic proposals designed at the GUC, and that would remained valid for the tropical regions until the late 1960s. The programs distinguished the professionals' activities of the employees, as well as their category, and established functional and aesthetic guidelines in accordance with these two principles.
iii. The third program handled by the GUC and that was identified as strategic in Caetano's Decree was the design of Urban Plans.

The importance of urban plans was based on the idea of network that its conception presupposes. This was reflected in an operation logic that was not only regional (of articulation of the different plans in a territorial level), but essentially urbanistic (resulting of the city's organization, observed as a typological model), In the urban layouts, much dependent of the technocratic thought of the architect João Aguiar (the main urban ideologist of the GUC), it was patent the definition of a set of functions that reproduced the responsibilities of each actor in the construction of the colonial city. Of private promotion would be the housing for the middle-class, and the secondary and tertiary sectors; while the State was responsible for the employees' housing, the educational, health and administrative facilities, later joined by the "traditional" housing for the African populations. Sports and leisure activities were shared (cinemas and clubs belong to the private initiative; on the contrary, stadiums were of public promotion).

The intention of these plans – whose existence as a project superimposed its application – was to consolidate the progressive image that the Estado Novo aspired to from 1945 onwards and that were truly, it believed, would ensure the future of its permanence in Africa. The urban plans were a tool that therefore functions in two levels: a) in the imaginary plan (giving clues about the colonial city that was formally desired); b) in the pragmatic plan (supplying indications on zoning, road layout and facilities to develop).

On these plans the standard buildings were placed, and were replicated in every city, potentiating (as it was expected) feelings of strong identification with the colonizing culture. This process of appropriation that alluded to a stereotyped idea of "Motherland" it was generally inherited of the first public buildings promoted by the regime even before World War II.

5 A CASE STUDY – SÃO TOMÉ AND PRINCIPE

The control of the colonial territory was also exposed in visions which envisaged the transformation of the landscape through a network effect, even if only in the imaginary level. Concerted actions, such as the one made by the Overseas Ministry when it commissioned in 1955 new urban plans for São Tomé and Principe settlements, reinforced the virtual dimension of many of the metropolitan proposals. The approach methodology started from the "existent population centers, to regroup the population, starting with the construction of elements which more directly serve the community" as reinforce the engineer Santos Paiva, the main technician responsible for the design plans (Paiva, [195–]: 2).

This proposed the replacement of the informal settlements that spread along the main roadway in low density residential neighborhoods, privileging single-family houses (lots with garden or vegetable garden) organized in orthogonal layouts and benefiting, when possible, of the preexisting matrix. The facilities were grouped in "civic centers", providing a bureaucratic nucleus to the settlements.

Of the design plans, only the facilities were built, but with a new location. Among the standard projects it can be find today the primary school, the sanitary post, with a modular structure in accordance with the number of inhabitants, the market and the switching station which provides electrical supply. A network of public fountains completed the sanitation process. These facilities function as urbanizing means, contributed to "create the interest in certain necessities of hygiene as well as others of the civilized life" (Paia, [195–]: 2). The proposed civilization model was still the European one.

6 PARTIAL FINAL CONSIDERATIONS

The designs that have been produced within the framework of Portuguese colonial policies have often had the objective of creating a collective consciousness about an imagined and not (exactly) concretized territory.

Many of the design drawings were then valued for their existence as "intention" and less for their operability. However, its production has allowed important decisions to be made as to how the territory was being infrastructures and the very nature of public facilities.

Figure 1. Low-income Housing for São Tomé e Príncipe, Alfredo da Silva e Castro, DGOPC-DSUH, trab.706, 1961 [AHU].

REFERENCES

Bonito, J. 2011. Arquitectura Moderna na África Lusófona. Recepção e Difusão das Ideias Modernasem Angola e Moçambique. Lisboa: IST – UTL.

Fernandes, J.M. (2002), Geração Africana – Arquitectura e Cidades em Angola e Moçambique, 1925–1975. Lisboa: Livros Horizonte.

Fernandes, J.M. et al. 2008. Moçambique 1875/1975 Cidades, Território e Arquitecturas. Lisboa: Edição de Autor.

Fernandes, J.M. et al. 2010. Angola no Século XX, Cidades, Território e Arquitecturas (1925–1975). Lisboa: Edição de Autor.

Fernandes, J.M. et al. 2013. São Tomé e Príncipe, Cidades, Território e Arquitecturas. Lisboa: Edição de Autor.

Ferreira, A.F. 2008. Obras Públicas em Moçambique – inventário da produção arquitectónica executada entre 1933 e 1961. Lisboa: Edições Universitárias Lusófonas.

Fonte, M.M. 2012. Urbanismo e Arquitectura em Angola – de Norton de Matos à Revolução. Lisboa: Caleidoscópio [2007].

Henriques, I.C.; Vieira, M.P. 2013. "Cidades em Angola: construções coloniais e reinvenções africanas", in N. Domingos; E. Peralta (orgs.), Cidade e Império – Dinâmicas coloniais e reconfigurações pós-coloniais. Lisboa: Edições 70, 7–58.

Magalhães, A.; Gonçalves, I. 2009. Moderno Tropical – Arquitectura em Angola e Moçambique 1948–1974. Lisboa: Tinta da China.

Matos, M. 2010. "Colonial Architecture and Amnesia", in J. Lagae; T. Avermaete, L'Afrique, c'est Chic. Architecture and Planning in Africa 1950–1970. Rotterdam: NAi Publishers, 25–34.

Mattoso, J. (dir.) 2010. Património de Origem Portuguesa no Mundo – arquitectura e urbanismo. Lisboa: Fundação Calouste Gulbenkian.

Milheiro, A.V. 2011. "Fazer Escola: a arquitectura pública do Gabinete de Urbanização Colonial para Luanda", in R.G. Prado; P.N. MARTÍ, La Modernidad ignorada – arquitectura moderna em Luanda. Angola, Madrid: Universidad Alcalá, 98–131.

Milheiro, A.V. 2012. "Africanity and Colonial Architecture: the Response of the Colonial Planning Office (1944–1974)", in EAHN, European Architectural History Network, Session: Regionalism redivivus. Do we need a closer look?, 326–330.

Milheiro, A.V. 2013. "The Republican city in the former Portuguese Africa", in Conference Colonial and Postcolonial Urban Planning in Africa. Lisbon: IPHS/IGSP-UL, 5–6 Sep.

Milheiro, A.V. 2017. African Colonial Architectures at the end of the "Portuguese Empire". Lisboa: Relógio d'Água.

Miranda, E. 2013. Liberdade & ortodoxia: infraestruturas de arquitectura moderna em Moçambique (1951–1964). Guimarães: Universidade do Minho.

Morais, J.S. 2001. Maputo, Património da Estrutura e Forma Urbana, Topologia do Lugar. Lisboa: Livros Horizonte.

Morais, J.S. 2010. Mindelo – Património Urbano e Arquitectónico – assentamento urbano e os seus protagonistas. Lisboa: Caleidoscópio.

Paiva, S. [195-]. Província de S. Tomé e Príncipe – Urbanização de Povoações. [AHU: cota OP08130]

Raposo, I. (org.) 200.. UR, Cadernos da Faculdade de Arquitectura da UTL – "Cidades Africanas", no 5 (Maio).

Tostões, A. (ed.) 2013. Arquitectura Moderna em África: Angola e Moçambique. Lisboa: ICIST, Instituto Superior Técnico, FCT [PTDC/AUR-AQI/103229/2008].

Drawing (...) City (...) Body, Dwelling on Earth – Janeiro (Ed.)
© 2019 Taylor & Francis Group, London, ISBN 978-0-367-02602-8

Drawing as a key feature in architectural creative process

A. Moreira da Silva
CIAUD, Lisbon School of Architecture, Universidade Lisboa, Lisbon, Portugal

ABSTRACT: As the architect must conceive and develop solutions for specific projects which can be of different nature, sketches may present themselves as an operative support for conceptual reflection, problem solving and critical analyses within the several phases of the architectural design process.

Through the study of several statements from various authors we search to investigate the permanence of drawings important role. From this theoretical approach we also intend to verify the relevance of drawing teaching in the formation of the future architects, despite the paradigm changes that emerge from times changes.

This research is based on a survey methodology with two different moments, being the first one an inquiry based on a questionnaire applied to undergraduate students, and the second a group of semi-structured interviews applied to architects. We intend to verify the permanence and the importance of hand-drawing in their daily design work, although the new assistive technologies constant progress.

We search to find fundaments that can prove that the fact of nowadays wide use of new technologies doesn't invalidate the important role played by hand-drawing as a stimulating instrument when sketching the first ideas and as a critical verification of the several hypotheses.

Keywords: Drawing, Architecture, Sketching, Creative process, New Technologies

1 INTRODUCTION

This paper stems from a current post-doc research project motivated for the need of producing more knowledge and reflection on Drawing and Drawing teaching. Our main research question is about the importance of hand-drawing as a critical instrument and as an operative support within the conceptual process in architecture and design.

The main objectives of this current research are: to verify the relevance of the hand-drawing teaching in the formation of the future architects, and also, to verify the permanence and the importance of hand-drawing in architect's actual daily work, although the new assistive technologies constant progress.

We aim to achieve contribution on the study of creative architectural process, enhancing hand-drawing important value for architecture students and professionals.

2 SKETCHING IN ARCHITECTURE

Sketching has been used for a long time, especially since Italian Renaissance when drawing started to assume a more important role as the basis of creation. In this period drawing assumed its intellectual character and some authors of those times, like Alberti or Vasari, considered it like something that the hand achieves coming from the human brain. (Moreira da Silva, 2010).

Since then we can find sketches as the basis of creative production not only in the fine arts but also in architectural projects. We can consider the relevance of hand-drawing in the act of designing not only as stimulating instrument when sketching the first ideas but also as critical verification of the several hypotheses.

Drawings have been used for generations of architects along several centuries. The use of sketches not only helps a great deal in starting the ideas generating process, it also enables high quality concepts to begin with as initial ideas. These ideas can be accessed, combined, selected and developed.

Drawing can be the key instrument of creation and control, also playing a critical role within the creative process in architecture.

Cross (2005) states that "drawings are a key feature of the design process. At the early stages of the projects they are communications with oneself, a kind of thinking aloud" and concludes that the conceptual thinking processes are based on the development of ideas through their external expression in sketches.

Sketching practice goes beyond the simple representation of ideas, it covers the development of intelligence from the perception and analysis of one's own drawing, a medium that is the reflection in action. (Schön, 2000).

In The Inexorable Rise of Drawing (2008), an article that analyses contemporary thinking about drawing, Annabel Tilley trays to identify the reasons why, nowadays, the drawing subject became so relevant. According to Tilley, "Drawing interdisciplinary nature has been acknowledge and also recognized drawing's facility to allude to or to describe the intangible way of articulating the inarticulate."

Several authors have written about the importance of drawing in architecture and design methodologies. Their general conclusion points out to drawing as an essential tool as it allows ideas to flow and provides critical investigation of the several alternative solutions in creative processes.

Given the difficulties of the project process as an ability to imagine what doesn't exist, drawing appears to be the project tool that enables the idea to become visible. Drawing mediates between thought and action, expresses the thought by images that allow us to visualize the future conception.

To study drawing, understood as a bearer and an intermediate of the idea and of the project, means to deepen its values in order to explore the multiple meanings contained in graphic language, understanding, directly, through the "reading" of the drawing, the generating idea of the project.

Architecture is inseparable from project methodology, and drawing assumes itself as an indispensable tool in the architecture process. The issue focuses on the relationship between drawing and project, and the importance that drawing acquires in the conceptual act of the project, as a booster instrument when registering the first ideas, and as verification of the several hypotheses.

Being aware of the architecture complexity as an intellectual ability for imagining what doesn't exist yet, drawing appears as the project essential tool that enables the possibility of making the idea visible. As the architect has to create and develop solutions for precise problems which may be of very different nature, drawing presents itself as an operative support for problems solving in the design conceptual process. Due to its operative characteristics, the importance of drawing assumes itself as a wide sense, providing the act of drawing

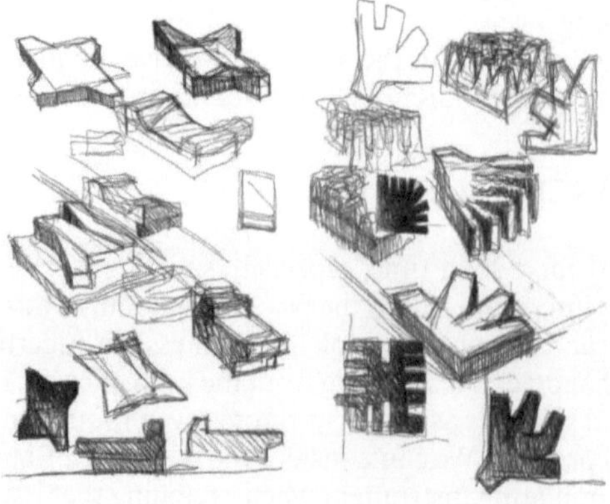

Figure 1. Sketches for architectural forms. João Santa-Rita (2010).

with the ability of being a means with multiple resources and a basis for the conceptual process underlying the practice of architecture.

Rohde (2011) states that "sketching is the visual thinking power tool" and adding sketching to the process is a great way to amplify software and hardware tools. He concludes that the use of sketching during the creative process, reveals the central role drawing plays within the development of a project and the importance of hand-drawing, particularly in an age dominated by digital media. For Rohde, sketches can work like a visual thinking tool and as a primary language for capturing thoughts, exploring ideas, and sharing those ideas. "Sketching provides a unique space that can help you think differently, generate a variety of ideas quickly, explore alternatives with less risk, and encourage constructive discussions with colleagues and clients."

Ortega and Weihermann (2017) stated that when architecture students use sketches, new ideas flow and qualities and relationships that were not previously imagined reveal themselves.

For Wilkinson (2016), it can be a simple drawing to start, but something happens when sketching that encourages the creative thinking and ideas start to flow, something begins to take form and "thinking through drawing" allows the evolution of several solutions and their critical analyses.

According to Norman (2010), the recent profound changes in project processes, due to the contemporary historical context were new technologies are so important, justify an enlarged discussion on the important role played by drawing in architecture courses, adapted to the new paradigm.

The fact of using new technologies doesn't invalidate the important role played by hand-drawing, both at the initial stage of recording the first ideas and during their subsequent development and in the critical analysis of the different hypotheses.

To Roome (2011) it is the possibility for cross fertilization between traditional and digital platforms that is interesting and beneficial.

Frank Gehry is one of the contemporary architects who most uses new technologies to achieve his works; however, he always uses sketches to fully develop his creative ideas and find solutions for his architecture projects.

Despite being at the forefront of high-tech innovation in architecture, renowned architect Chris Wilkinson believes passionately in the importance of drawing by hand. Where many practices are now dominated by computer aided design, Wilkinson still uses drawing as a way to think through ideas, to grapple with design problems and as a tool of communication. (Pearman, 2015).

In a recent interview Wilkinson declares: "I think it is essential for architects to be able to draw and I wish it was encouraged in schools. I'm convinced that drawing and sketching is essential to the design process" and recognizes the importance of the daily use of his sketchbook: "I carry a sketchbook with me everywhere I go and it provides me with another way of thinking and resolving ideas in my head and provides a useful memory bank". (Wilkinson, 2014)

For Wilkinson (2016) the strength of the drawing lies in the way it is part of his thought process. About the use of drawings in his architectural practice, Wilkinson says: "I always start projects with a pencil on paper."

Norman Foster is also one of the contemporary famous architects who uses drawing on a daily basis. Willkinson (2014), recalling his young days in the studio of Norman Foster, says that he could consume a whole sketchbook at a sitting, drawing at speed, leaving a literal paper trail of ideas to his collaborators.

Figure 2. Sketch by Frank Gehry for Guggenheim Museum Abu Dhabi (2010).

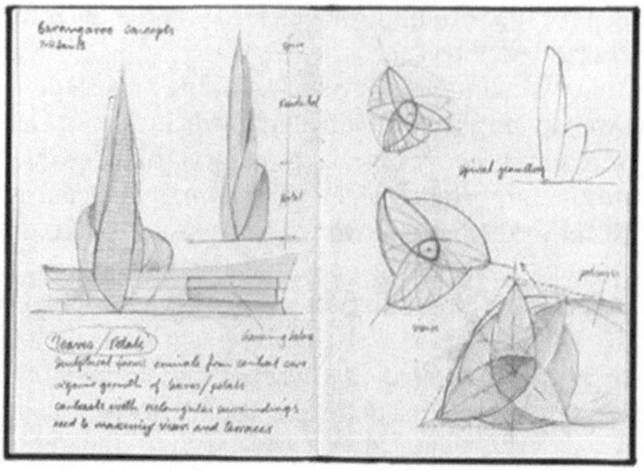

Figure 3. Wilkinson sketchbook with details of the plans for the Crown Hotel, Sydney, Australia (to be completed in 2019).

Figure 4. Norman Foster first sketches for London City Hall (2002).

The complexity, characteristic of the transition from the idea to its materialization, is related to the actual extent of architecture where a number of factors come into play, such as the ability to understand the context and to imagine the solutions, the ability to know and take advantage of the processes and materials as means or vehicles through which the solution is materialized, the ability to transform ideas into appropriate forms, having in mind the limits and material possibilities.

Within the complex process which takes place from how to imagine a project solution till its materialization, the architect can use hand-drawing as an indispensable mean to help him to develop the idea, as an operative support in the creative process. This attitude comes across the concept of drawing as a mental activity, which uses the hand as an extension of the brain. The search for solutions, even for simple problems, implies that drawing studies in detail each phase of the process for obtaining the result we seek for. The complexity of the required drawings emerges as the architect approaches the solutions which he considers ideal or that are possible.

During the conception phase we should not disregard the importance of a trace, of a scribble, of a sketch. They can all contribute to the idea development.

For an architect, the sketchbook is not just a place to draw but also a place to order thoughts, to graphically and visually gather information and to develop a design response. It merges as a way of thinking through drawn lines as a creative process, revealing the central role that drawing can play within the development of any project.

"The real goal of sketching is about generating ideas, solving problems, and communicating ideas more effectively with others." (Rohde, 2011).

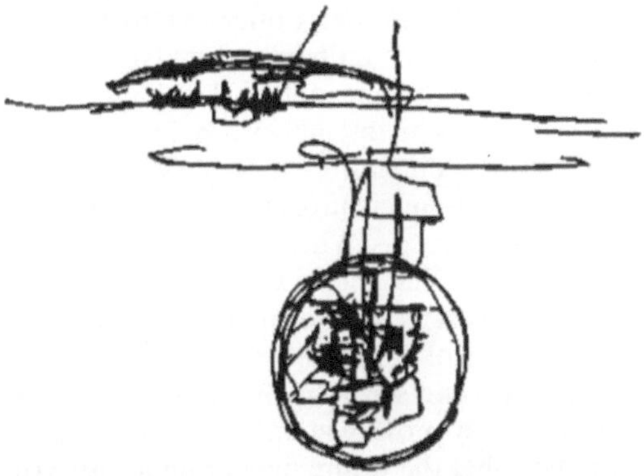

Figure 5. Jean Nouvel sketches for the Louvre Abu Dhabi (2010).

The act of drawing allows that the reasoning and thoughts we have developed can be gradually translated and decoded throughout the drawn lines. Somehow, we debate ourselves with our own ideas in the paper space. We scratch, we draw, we overwrite features, we configure, we represent, we visualize, giving physical form to our thinking.

There is a direct link between the thinking and the hand that performs the drawing: the hand as an extension of the brain, of the reasoning.

For Hamilton (2009) the value of hand-drawing is associated with the knowledge that drawing is the essence of creativity and its measure would lie in the individual ability to continuously express and create, using self-exploration and creative thinking through the process of drawing, linking together interesting ideas and technical know-how, enhanced by assistance and interaction with new technologies.

Domingos (2017) in his recent book, writes about the extraordinary advances of artificial intelligence and its promising future, but as he mentioned throughout this book, artificial intelligence is developing in the direction of automatic learning, but not yet capable of creating something real new which is the very essence of human creativity.

Drawing holds, in itself, the making visible quality of the whole mental process underlying the conceiving stage, from the first sketches of a project vague idea to the creation of its final form.

Drawing's transcendental importance lies in the capacity that it gives the architect to materialize abstract conceptualizations and to create the ideational basis for architectural forms and buildings.

According to the French architect Jean Nouvel, his latest work, the Louvre Abu Dhabi, recently opened, began to take shape as a simple sketch done over a paper napkin during a meal in a restaurant. This is an example of the versatility of free hand-drawing in terms of use: we can draw on any support and with any media, anytime and anywhere.

Drawing today serves as a primary medium for generating, testing and recording an architect own creative and conceptual musings about a project.

If drawing is central for conceiving, it is also central for defining how that conception is managed as it moves from its initial stages through its actual development and realization as a material form. Drawings illustrate various aspects of the architectural creative process. So, the importance of drawing assumes a broad sense, conferring to the act of drawing the ability to become a means of multiple resources to practice the discipline of architecture.

3 CONCLUSIONS

From many authors' statements and according to the professional practice of many architects, we can conclude, from this research initial phases, that drawings are the key of the creative process in architecture. Drawing presents itself as an operative support for problem solving and critical analyses in the architectural creative process.

We can also conclude that new technologies offer the architect new ways, allowing him to save time and facilitating many of the daily practice tasks, even becoming indispensable in many work phases. However, they do not replace sketching, which continues to assume a 'chameleon' shape during the various historical periods, in a constant time adaptation, and, especially by incorporating a critical dimension in design's creative process.

New technologies can be a stimulus for creative capacities, but it doesn't invalidate the important role of sketching in the creative process. When executing a drawing, the one who executes it makes choices, composes, and selects what and how to communicate, both in the representation of reality (objectivity/mimesis), and in its transformation (interpretation/invention).

Drawings can show appearances of things as experienced by someone who draws, evoke some aspects of human existence, and seize the meaning of the subject under inquiry. The experience of freehand drawing may provide attentive awareness and add reflexive value.

Although we cannot predict what the future may bring in the Artificial Intelligence field, we believe that the creative capacity that sketching can trigger will hardly be replaced by thinking machines in a near future.

Despite the paradigm shift required by the changing times, we believe that hand-drawing will remain inseparable from the architects training and professional practice, assuming an essential operating support in the project activity which continues being the architecture basis.

In the future, design education should pass through a systematic approach to hand-drawing in order to highlight and analyze the flexibility with which it adapts itself to various purposes, fulfilling effectively a wide range of intentions, in a constant and vital adaptation to the continuous changes in the teaching and in the practice of the student and of the architect, given the new techniques and technologies.

In order to validate these assumptions, the current post-doc research can constitute a contribution for the understanding of the importance of hand-drawing's permanence as the basis of the architectural creative process, although the constant progress of the new technologies.

REFERENCES

Cross, N. 2005. Engineering Design Methods, Strategies for Design. John Wiley & Sons Ltd. Chichester, England.

Domingos, P. 2017. A Revolução do Algoritmo Mestre. Manuscrito Editora, Lisboa.

Hamilton, P. 2009. Drawing with printmaking technology in a digital age. http://www.lboro.ac.uk/microsites/sota/tracey/journal/dat/images/paul_hamilton.pdf.

Moreira da Silva, A. 2010. De Sansedoni a Vasari: um contributo para o estudo do Desenho como fundamento do processo conceptual em Arquitectura. Universidade Lusíada Editora. Lisboa.

Norman, D. 2010. Why Design Education must change. Retrieved from: http://www.core77.com/blog/columns/why_design_education_must_change_17993.asp.

Nouvel, Jean. Sketch retrieved from: https://www.capitalfm.co.ke/news/2017/11/jean-nouvel-napkin-sketch-abu-dhabis-architectural-triumph/.

Ortega, A. R., and Weihermann, S. 2017. Graphic dialogues: The progress of knowledge in architecture studio. In: Progress(es) – Theories and Practices. Kong, M.S.M. and Monteiro, M.R. (eds), pp. 111–114. CRC Press, Taylor & Francis Group, London, UK.

Pearman, H. 2015. Thoughts on paper: the sketchbooks of Chris Wilkinson RA. Retrieved from: https://www.royalacademy.org.uk/article/thoughts-on-paper-chris-wilkinson-ra.

Rohde, M. 2011. Sketching: the Visual Thinking Power Tool. Retrieved from: http://alistapart.com/article/sketching-the-visual-thinking-power-tool.

Roome, J. 2011. Digital Drawing and the Creative Process. Retrieved from: http://www.lboro.ac.uk/departments/ac/tracey/tracey@lboro.ac.uk.

Schön, D. 2000. Educando o profissional reflexivo: um novo design para o ensino e a aprendizagem, Artes Médicas Sul, Porto Alegre.

Tilley, A. 2008. The Inexorable Rise of Drawing. Retrieved from: http://www.transitiongallery.co.uk.

Wilkinson, C, 2014. Interview. Retrieved from: https://www.designboom.com/architecture/interview-with-chris-wilkinson-founder-of-wilkinson-eyre-architects-08-05-2014/.

Wilkinson, C. 2016. Thinking through drawing. Retrieved from: https://vimeo.com/159085592.

Drawing (…) City (…) Body, Dwelling on Earth – Janeiro (Ed.)
© 2019 Taylor & Francis Group, London, ISBN 978-0-367-02602-8

The thickness of the world

D. Loução & M. Cruz
Lisbon School of Architecture, Universidade Lisboa, Lisbon, Portugal

ABSTRACT: Inspired in Mitra, an architectural work conceived by the Portuguese architect Vitor Figueiredo (1929–2004), we approach the constitution of space by the body – phenomenological architecture. We explore a dwelling incarnated in the object, a corporeal phenomenon – pure experience. In other words, a communication between object and subject, and in that way ask for a Pure Architecture.

In the first section we analyse the architect's attitude as a phenomenological movement, based on the notions of 'absolute givenness' and 'pure phenomena', from Husserl (1859–1938) and the contributions of others philosophers as Merleau-Ponty (1901–1961), and Lyotard (1924–1998), aiming to achieve in the [architectural] object an immanent knowledge through the experience it. In the second section, drawing from the idea of experience by dwelling in the world and with the contributions of Merleau-Ponty on 'perception' and the 'body', and the insight of the theoretical architect Juhani Pallasmaa (1936–), we explore the embodied experience of the body by being in the world and the body as an expression of the world, in order to reveal a corporeal dwelling founded by the nature of being with the world, in other words, the dwelling in experience [Intentionality]. In experiencing Mitra, we found a new kind of dwelling and of making Architecture. The work is believed to reveal the incarnate habit that the architect-author himself lived (experienced/felt), that is, the corporeal value under the symbolic value, materialized in the work by the architect and inhabited in the experience of the user subject.

Keywords: Phenomenology, Experience, Pure Architecture, Vitor Figueiredo

1 INTRODUCTION

Mitra[1] speaks for itself: "[…] had we not eyes, or more generally senses, there would be no painting [architecture] at all for us, yet the picture [object] 'tells' us more than the mere use of our senses can ever do." (Merleau-Ponty, 2005: 347). But how can the architectural object – 'invariable structure' (Merleau-Ponty, 1995: 78) – speak for itself? Although it is constituted by limits [elements], as Schulz understands it: "The boundaries of a built space are the floor, the wall and the ceiling." (Schulz in Nesbit, 2005: 450), because otherwise it would be impossible to make itself present, the Mitra – architectural object – is also something else, without even being able to speak in matter or the result of it – the space. Can Architecture be, something else, beyond the object or the space in it, or in Husserl's words, how can we see it as an 'absolute givenness' (Husserl, 1999: 26)?

> "[P]hilosophy [Architecture] must reflect on the mode of presence of the object to the subject, on the conception of the object and the conception of the subject as they are shown in the phenomenological revelation [attitude], instead of replacing

1. "Mitra is a designation used by Vitor Figueiredo and his collaborators to refer to the project and work of the Herdade da Mitra Pole Agronomy complex, part of the University of Évora" (Arenga, 2012: 38). Won in competition in 1992 and built in 1995.

Figures 1 and 2. The first image is the impact on the arriving street and the second image evinces the space between the volumes of the buildings. Illustrations are written by José Maria Assis e Santos in O detalhe do projeto geral, a sombra: O polo da Mitra de Vítor Figueiredo.

> them by the relation of the object with the subject as it is conceived in an idealist philosophy [Architecture] of total reflection." (Merleau-Ponty, 1999: 99).

2 PHENOMENOLOGY

2.1 *The attitude as phenomenological reduction*

Like Husserl's assistant, the philosopher Eugen Fink (1905–1975), spoke of a 'wonder' befor the world (Ibid. xiii) to explain "the best formulation of the reduction" (Merleau-Ponty, 1995: xiii). The architect Vitor Figueiredo (1929–2004), speaks about 'curiosity' as a way to define an 'architecture of faith', instead of an 'architecture of superstition': "[…] one in which each person tries […] searches his way; more than make up your career, try to do your thing!" (Ibid. 54).

"What differentiates us [...] is curiosity" (Figueiredo in Arenga, 2012: 54). The 'curiosity' of the architect defines the architect's attitude – a phenomenological one – not only the way the architects should work "...question the world" (Ibid. 54), but also the way as he sees the profession. To be an architect – by attitude – is to be curious about the world, to suspend all the knowledge already given. In that sense, one ought to suspend not only the scientific knowledge but also the reality of the world, i.e. the belief in the reality of ordinary experience: "the experience of the world as the true universal ground of knowledge" (Husserl, 1975: 6). Furthermore, to be an architect – by profession – is a deliver in the act of practice, that is a delivery in the things themselves, the things each one gives to himself, as the architect affirms.

Delivery doesn't mean submission, or deprivation, instead it means 'pleasure' – 'enjoyment' to use Figueiredo's expression – the pleasure of delivery in the profession, or in other words, the pleasure of delivery in the things... It could be understand as the way things are made present to me and by that, an act of delivery in the world because the way each one gives himself/herself, in the act of practice, in order to grow as a person.

Figueiredo's 'curiosity' must signify not only the act of suspending the belief of reality of the world of ordinary experience, but simultaneously, an act of delivery in the world, as the way as the architect understands the profession. We would like emphasize here, the apparent contradictory meaning of the word 'curiosity'. It is in that sense that we selected the word of Fink – 'wonder' – to explain the method of reduction used by architects. Reduction as we have shown is normally identified as to "put out of play" (Lyotard, 1991: 47), but it also means that we must remain with the things...

> "One must not go beyond the piece of wax in doing a philosophy of extended substance, nor in doing a philosophy of the a priori spatial forms of sensibility;

one must remain with the piece of wax itself, describe only what is given, without presuppositions." (Lyotard, 1991: 33).

To better explain the architect's attitude as 'wonder' before the world, we can follow the words of the theoretical architect Juhani Pallasmaa, in the article: "Space, Place and Atmosphere – peripheral perception existential experience" (2011) published in the magazine Lebenswelt, when he says: "The all-encompassing and instantaneous perception of atmospheres calls for a specific manner of perception – unconscious and unfocused peripheral perception". The author states that we live automatically, without a permanent, active and systematic attention of what surrounds us, that is, we inhabit by means of a 'peripheral perception'. We can say that the 'peripheral perception' of Pallasmaa is an exploration of Husserl's act of seeing. It clarifies the apparent contradictory meaning in it, because it means to see as peripheral (not direct vision) and in this sense as a suspension of the reality, but at the same time to see (direct vision), as delivery in the things. So one has to keep the distance[2] from the distance that reduction claims, because it is in seeing of this distance that we can learn to see the wonder before the world, and that is what Figueiredo means by 'curiosity'. But in what way can the architects create such distance in order to obtain such knowledge? Or, how can they see such knowledge? Or following Lyotard in the questions of Husserl: "[H]ow real experience enters the a priori realm of all possible knowledge"? (Lyotard, 1999: 44)

2.2 *The pure ego as the transcendental experience*

Instead of the belief on the reality of the world through ordinary experience, where "I discover [reality] as existing and receive it, as it gives itself to me, equally as existing" (Husserl in Lyotard, 1991: 46), admitting a world as real or as unreal, where "Corporeal objects are simply there for me in some spatial distribution; they are 'present', in a literal or figurative sense [...]" (Ibid. 45) –, one reality that "assimilates the cogito to an axiom of knowledge" (Ibid. 47), in other words, a world raised by logical intuition in natural attitude, we claim the reality of the world, by the 'curiosity' of the architects – phenomenological reduction –, a world raised by sensible intuition from where "I take no position with respect to the world as existing" (ibid. 47), only "given to me only through sensory experience" (Husserl, 1975: 7) then the world for me is only possible through phenomena and "what appears valid in such cogitationes." (Ibid. 8), in Husserl words: "the surrounding world is no longer simply existing, but "phenomena of being" (Husserl in Lyotard, 1991: 47)

> "Ordinary experience draws a clear distinction between sense experience and judgement. It sees judgement as the taking of a stand, as an effort to know something which shall be valid for myself every moment of my life, and equally 'for other actual or potential minds; sense experience, on the contrary, is taking appearance .at its face value, without trying to possess it and learn its truth." (Merleau-Ponty, 1995: 30).

As Lyotard (1991: 45) states, Husserl's admiration of Descartes is his method of doubting, in particular, the recognition that the locus of truth lies in the subject of knowledge. However, and as differentiation between both, is not the isolated subject, instead the opened [reduced] subject to the world, and maybe world and Merleau-Ponty speaks in being-in-the-world. In this way, the architect's 'curiosity' as a phenomenological attitude is because it does not isolate the natural self, it does not make the "ego to a substantia cogitans, to an independent human animus" (Husserl, 1975: 9) – the Cartesian Ego –, but reduces it to the transcendental self. Instead of "using the ego cogito [...] as an absolutely primitive premise" (Husserl, 1975: 11), Husserl makes the shit to the "ego cogito, as the apodictically certain and last basis for judgment" (Ibid. 7) and as Merleau-Ponty (1995: xiii) mentions, it allows reflection not to

2. The word 'distance' is used acknowledge the double meaning of it: "...there is a distance which separates me from life or, rather, which unites me with life" (Minkowski: 1970: p. 403).

withdraw from the world "[...] it steps back to watch the forms of transcendence [...] conceiving the subject as a process of transcendence towards the world."

Looking for knowledge that provides for itself, architect's 'curiosity' suspend the reality of the world in ordinary experience, which means, losing the "entire concrete world" (Husserl, 1975: 7) and stop accepting as real or unreal. This suspension allow us to pay a tribute to Descartes: "The methodology of a purified Cartesianism demands of me, the one who philosophizes, the phenomenological epochi" (Ibid. 10), but at the same time claims for an evidence [of phenomena – the cogito] by sensory experience, where the world rendered possible as a phenomena [experience], regardless of the truth it claims for the being in the presence of the phenomenon, since we cannot deny the evidence of phenomena, in other words, we cannot affirm an impossibility of phenomena. We would like to affirm that "epistemological abstention does not mean their disappearance, at least not as pure phenomena" (Ibid. 8) and that is the main discover that Husserl made: "To return to things themselves" (Merleau-Ponty, 1995: ix).

We would like to on focus this dangerous point as 'the abysm', because it is in this position that the architect should posited himself if we want see the world, in others words "to understand myself [himself] as that ego and life of consciousness in which and through which the entire objective world exists for me" (Ibid. 8). This means that 'curiosity' should have implicit the experience [evidence of cogito], because the world for me is "what I aware of and what appears valid in such cogitationes." (Husserl, 1975: 8). In other words, Figueiredo's 'curiosity' sees "life exclusively as consciousness of the world" (Ibid. 8), the architect's 'curiosity' [architect's attitude] shouldn't step away from the world, instead they must focus in the world to see these life as piece of the world and how the world is also piece of this life, revealing the pure Ego:

> "I cannot live, experience, think, value, and act in any world which is not in some sense in me, and derives its meaning and truth from me." (Husserl, 1975: 8).

The World of Phenomena as Sphere of Absolutely Indubitable Givens Lyotard (1991: 46) as Husserl (Ideal of Phenomenology 1907 and in Ideas I), recognizes in the Cartesian inspiration the 'concrete subject'. Certainty the experience of doubting is the guarantee of the act of doubt when we're doing it: "in making the judgment that everything is doubtful it cannot be doubted that I am making this judgment" (Husserl, 1999: 23), and the same is true, as the author recognizes, for every cogitatio – perceive, imagine, judge, infer: "...it remains absolutely clear and certain that with respect to perception I am perceiving this or that, that with respect to judgment, I am judging this or that, etc." (Husserl, 1999: 24). These acts – real or unreal, objects or not – are the "title for a highly ramified sphere of being" (Ibid. 24), what Husserl identifies as knowledge, and can be given to us absolutely, as long as we "reflect on them, accept them and posit them in a pure act of seeing" (Ibid. 24).

> "Every intellectual experience, indeed every experience whatsoever, can be made into an object of pure seeing and apprehension while it is occurring. And in this act of seeing it is an absolute givenness. It is given as an existing entity, as a "this-here". It would make no sense at all to doubt its being." (Husserl, 1999: 24).

Husserl replaces inner perception – Cartesian perception –, with reflective vision which is replacing thought by experience. That is what we mean by experience a "giveness in the field of ego cogito" (Husserl, 1975: 9), by the 'act of seeing', in Lyotard words, is not "a simple representation [...] symbolic comprehension" (1991: 44), but "sensible intuition" (Ibid. 44) of experience: "the intuition of experience by itself constitutes the model of all originary evidence" (Ibid. 45). So, as said above, following Husserl, Lyotard (1991: 45) recognizes in the Cartesian subject on the act of doubting the 'concrete subject' – an absolute, a 'this'. This 'absolute' is not an abstract framework but instead a 'lived reality', in our words, it's not a relation [metric] with something but instead a presence[3] [alive] of something to my body: "the

3. We warn the reader that the meaning of the expression is not intended to affirm a palpable [physical] reality: "Such a person could not even imagine them, that is, could not see them and in seeing them apprehend how they are possible" (Husserl, 1999: 30).

perception, as long as it lasts, is and remains an absolute entity, a "this-here" (Husserl, 1999: 24), in others words, a genuine experience absolutely given in the self, and that is why we can say impossibility of not experience it. As Merleau-Ponty (1995: 191) mentions, an embodiment of experience is in a way similar to an illusion because you can't step way of it, of our body, understood this body as an atmosphere[4]. It is as Pallasmaa (2011: 39) affirms, when he speaks of multisensory experience, agreeing with the philosopher Merleau-Ponty (1995: 194): "Therefore there are not the senses, but only consciousness."

> "To say that it is still myself who conceive myself as situated in a body and furnished with five senses is clearly a purely verbal solution, since I who reflect cannot recognize myself in this embodied I, since therefore embodiment remains in the nature of the case an illusion, and since the possibility of this illusion remains incomprehensible." (Merleau-Ponty, 1995: p. 191).

These acts, while still happening – presently existing or in imagination –, they are an 'absolute entity', and that's the reason why we can speak of them, "not in vague and sketchy terms" (Husserl, 1999: 25), but rather, we can "examine their essence, their constitution, their imanent character and conform our talk by a pure measure to what is seen in the fullness of its clarity" (Ibid. 25), because they are objects of intuition[5], and so we would affirm experienced in the co-presence (lived or imagined) of my body.

> "…how does it come about that we perceive? We shall understand this only if the empirical self and the body are not immediately objects, in fact only if they never quite become objects, if there is a certain significance in saying that I can see the piece of wax with my eyes, and if correlatively the possibility of absence, the dimension of escape and freedom which reflection opens in the depths of our being, and which is called the transcendental Ego, are not initially given and are never absolutely acquired" (Merleau-Ponty, 1995: 186).

Seeking for the essence of knowledge, or as Husserl explains: "to bring the essence of knowledge to direct self-givenness" (1999: 25) Husserl refers the Cartesian doubt as the circle of absolute givenness: "which is comprehend under the title of the evidence of the cogitatio" (Ibid. 26). Experience as the 'evidence of cogitatio' does not aim at the movement of the body – physical or psychological experience – "It will not explain knowledge as a psychological fact" (Husserl, 1999: 25) –, but rather an incarnation of the body in things, in space: "To return to things themselves is to return to that world which precedes knowledge, of which knowledge always speaks" (Merleau-Ponty, 1995: ix). Experience is a phenomena, and recognizing the Cartesian inspiration – an impossibility to doubt –, assimilated in the body as an 'absolute givenness', as an entity, a 'this-here': "It is an absolute this-there, something whose existence it makes no sense to doubt" (Husserl, 2008, p.195).

So, this incarnation in experience as an 'absolute givenness' is assumed as an impossibility to doubt. When we say the impossibility to doubt we intend to identify the evidence of phenomenon between the body and the architectural object, and in this sense an 'absolute givenness' of experience – our idea of Architecture –, which by reflective vision makes possible the contact of the ego, opened by the phenomenological method – the Figueiredo's (and we hope of all architects) 'curiosity' – with this world:

> "But the reflective I differs from the unreflective at least in having been thematized, and what is given is not consciousness or pure being; it is, as Kant himself profoundly put it, experience, in other words the communication of a finite subject

4. "Atmosphere is something between the subject and the object" (Gernot Böhme *in* Borch, 2014: 43).
5. "When phenomenology examines objects of consciousness—regardless of what kind, whether real or ideal—it deals with these exclusively as objects of the immediate consciousness. The description—which attempts to grasp the concrete and rich phenomena of the *cogitationes*—must constantly glance back from the side of the object to the side of consciousness and pursue the general existing connections." (Husserl, 1975: 16).

with an opaque being from which it emerges but to which it remains committed" (Merleau-Ponty, 1995: 196).

We may say, an Architecture absolutely given in the self. Figueiredo's 'curiosity' – and by this method, his Architecture – does not seek the validity of ordinary experience – the being of the world – but rather to "[...] learn that the world and how the world is for me [him] the cogitatum of my [his] cogitationes" (Husserl, 1975: 15). The point here is to show how architecture can be something more than the necessary construction of the object or the result of it – the negative space. In other words, Architecture understood as an absolute giveness in the self – Pure Architecture –, something behind the physical matter of the object – solid or void –, and in that way, something that questions our existential condition, seeking for news ways to experience life, one that expresses what unit us, in this world, as Merleau-Ponty wrote it (1945), and as Pallasmaa mentions: "Architecture continues to have a great human task in mediating between the world and us, and in providing a horizon of understanding of our existential condition." (2017: p. 75)

3 EXPERIENCE EMBODIED PRESENCE

In the book The Philosophy of Walking (2013), the author Frédéric Gros (1965–), presents the philosophy of 'walking': "when you are walking, there is only one sort of performance that counts: the brilliance of the sky, the splendour of the landscape. Walking is not a sport" (Gros, 2014: 02). The verb[6] 'to walk' is used as a metaphor for the movement of the body, not only the body movement, but – and following the concept of delivery in architect's attitude – the movement of the body in [towards the] world, or in other words, the experience in the world – the being of cogitatio made possible by the sphere of cogitations (Husserl, 1999: 33). Following the ideas of Merleau-Ponty (1995: 71) about the body as a medium to be conscious of the world: "the body is the vehicle of being in the world, and having a body is [...] to be intervolved in a definite environment" and Otto Friedrich Bollnow (1903–1991) in his book Human Space (1963) where he presents the wandering by an observation of Linschoten: "... anyone who twists his ankle while walking has not properly understood the countryside" (Linschoten in Bollnow, 2011: 109), to the point of concluding that the landscape is not an image, like the modern road, but it is part of the traveller: "[the wander] is completely taken up by it [countryside]" (Bollnow, 2011: 109); the author, Frédéric Gros, quoting the philosopher Friedrich Nietzsche (1844–1900) seeks to demonstrate that, not only, the experience of the world reveals the body, but also, that the experience of the body seizes the world:

> "We do not belong to those who have the ideas only among books [...] It is our habit to think outdoors – walking, leaping, climbing, dancing, preferably on lonely mountains or near the sea where even the trails become thoughtful..." (Nietzsche in Gross, 2012: p.18).

The main subject, is to be able, to understand the relation between our body – usually seen as an object or seen as another person besides the self – and experience, as George Lakoff (1941–) and Mark Johnson (1949–) in their book Philosophy in the Flesh (1999) write about the 'embodied mind', inspired by the ideas of Merleau-Ponty on the embodied experience of the felted world by living on it, and the idea of John Dewey (1859–1952) about how the bodily experience is connected with the things that we can think, know and communicate (Lakoff and Johnson, 1999: 9). In other words, the meaning of the word experience, that which links the object to consciousness: "...consciousness is always consciousness of something – that is, that consciousness is intentionality [...] the manner in which I know the object, and in which the object exists for me" (Lyotard, 1991: p. 43) as the expression of Merleau-Ponty (1995: 21):

6. In opposition to Schultz who classifies space with *nouns* (Schulz *in* Nesbitt: 2006: 452), Pallasmma prefers the *verbs* (Pallasmaa: 2011: 59).

"[T]he 'human world' ceases to be a metaphor and becomes once more what it really is, the seat and as it were the homeland of our thoughts".

Gross's philosophy is the dwelling in experience, realized by the 'being-in-the-world' (Merleau-Ponty, 1995: 68), and which the philosopher calls the 'I committed': "...our inherence in a world..." (Ibid. 71), and by this dwelling, we mean intentionality – "the ego, like the situation, is definable only in and by this relationship" (Lyotard, 1991: 80) –, the opening to a knowledge in experience, and because of it, we may state a silent knowledge acquired by the body in experience by being in the world, and as Pallasmaa recognizes, the difficulty to apprehend it: "our corporeal existence is rarely identified as true interaction and integration with the world or our self-awareness or our conscience" (Pallasmaa, 2013: 12), we may call it: embodied knowledge. It is embodied because it is the committed body – not only the body as our vehicle, but also the body as expression of being in the world – that stores knowledge: "It is thought born of a movement" (Gross, 2012: 20), and because of this kind of experience, Pallasmaa (2013: 109) refers the embodied thinking, that is, thought born of the movement of the body in the world, which the every person, and, in particular, the architects, should use in the operation of making and dwelling. Is the operative act of the architect exclusive on the act of making or does the act of experience – dwelling – be part of the making? When architects experience a building, aren´t they making?

The dwelling in experience expresses the contact with the world: "my contact with ambient becoming is achieved across, or rather with the help of, a 'distance' which unites us to it." (Minkowski, 1970: 403), which according to Merleau-Ponty (1995: xvi) means the unity in the world: "recognizing consciousness itself as a project of the world, meant for a world which it neither embraces nor possesses, but towards which it is perpetually directed", the revelation of a corporeal existence in the act of dwelling as the expression of Lakoff and Johnson (1999: 15): "reason is shaped by the body", but in our words, we would affirm knowledge instead of reason: knowledge is shaped by the body, and as Lyotard (1991: 80) so well explains "the world is denied as exteriority and affirmed as "surrounding", the ego is denied as interiority and affirmed as "existing"", the dwelling who apprehends the thickness of the world, and understood through our body, the matter of the world:

> "The significance of the ash-tray (at least its total and individual significance, as this is given in perception) is not a certain idea of the ash-tray which co-ordinates its sensory aspects and is accessible to the understanding alone, it animates the ash-tray, and is self-evidently embodied in it. That is why we say that in perception the thing is given to us 'in person', or 'in the flesh'." (Merleau-Ponty, 1995: 287).

This phenomenon of being, wrapped in an act of seeing [feeling] and in a an act of moving [action], as the philosophy of Husserls and Merleau-Ponty explains, in other words, this phenomenological attitude, allows an internalization of experience [or rather phenomena] as Pallasmaa emphasizes: "...movement, balance and scale are sensed unconsciously throughout the body, such as tensions in the muscular system and in the positions of the skeleton and other organs." (2011: 63), and by that, we can speak of an embodiment presence: "self and world merge in the activity of dwelling, so that one cannot say where one ends and the other begins, it surely follows that the intentional presence of the perceiving agent, as a being-in-the-world, must also be an embodied presence" (Ingold, 2000: 169), expressed in Figueiredo's attitude: "There are times when I go in the plain of the Alentejo, I stop the car and take five to ten minutes, alone, with myself. There are times when I hardly look at the plain, although I love it deeply.

Because I cannot stand it", and evident in Mitra, as we will try to demonstrate following the words of Pallasmaa:

> "Buildings are not abstract constructions and without meaning or even aesthetic compositions, they are extensions of our bodies and shelters for them, as well as our memories, identities and minds. Consequently, architecture arises out of our existentially true confrontations, experiences, memories, and aspirations." (Pallasmaa, 2013: 119–120).

And so, about it – the Mitra – the architect would say: "We all felt that an objective situation [...] would not take place here. [...] We were worried that this was not a gloomy thing and that this space – without rigid, geometric rules, drawing, or this or that – would translate that very thing." We looked for a certain in Arenga, 2012: 48. Mitra is not an idea, says the architect many times, with Mitra, I come to [see] myself, corporeally questioning – with all my body –, the world. In seeing, I grasp each bulge of its surface in my hands, weight the mass of each column in my body, lengthen the presence of the shadow on my skin, and feel the rays of light piercing the space. In motion I seize things. Mitra is this sensitive being [myself] that allows the encounter with the world, through the body, not the mechanical body, but the perceptive being, lived by it – my body. Sunlight is not mechanically braked, but filtered [understand] under my gaze, small rays of light violate the now-present atmosphere in me. All this void [solid], which begins at the edge of my skin and ends in the palpable things, solidifies by the mistery of my sight and become a sensitive body, which I touch we all my body, touching me. Following Minkowski, we can say, there is distance, or in other words, there are no interstitial spaces... There is absolutely given in my body. I see the body of the world – the matter – by my own body as a seer and sentient: "There is another bet – which has nothing to do with it and has to do with faith – of a more real situation" (Figueiredo in Arenga, 2012: 55).

Instead of an idea [of the object], Mitra is the experience [of the subject], but not only the experience understood as the concrete movement in the world, but also the abstract movement made possible by the body as "the medium [. . .] the potential source of a certain number of familiar actions" (Ibid. 91), a 'human space' identified by a zone of reflection and subjectivity (Merleau-Ponty, 1995: 97), made possible by the act of seeing: "reflection is not absolutely transparent for itself, it is always given to itself in an experience" (Merleau-Ponty, 1995: 38). We do not see the object or the projections of it, and so we cannot speak of elevations, as we cannot speak of pillars, doors or windows – elements of architecture –, as Pallasmaa (2011: 60) inform us: "The elements of architecture are not visual or gestalt units; they are encounters, confrontations that interact with memory". The perception of space is not the positive or the negative of the object – partes extra partes (Merleau-Ponty, 1995: 63) –, but an extension of my body, understood in pure experience: "The sides [images] of the cube [object] are not projections of it, but precisely sides [absolutely givenness]" (Merleau-Ponty, 1995: 182). The perceptual experience is our body in the world and I cease to see the objective world – elevations (objects) – and I experience the world as phenomenon – images-data (absolute givenness) as expressed by Figueiredo's statement: "There are no interstitial spaces [...] The space between buildings [my space] is the generator of architecture itself." (Figueiredo in Arenga, 2012: 48).

> "But in the concrete act of reflection, I abolish this distance, I prove by that very token that I am capable of knowing what I was perceiving, I control in practice the discontinuity of the two selves, and it would seem that, in the last resort, the significance of the cogito lies not in revealing a universal constituting force or in reducing perception to intellection, but in establishing the fact of reflection which both pierces and sustains the opacity of perception." (Merleau-Ponty, 1995: 38).

And that is why we can speak of the between: "if the words 'enclose' and 'between' have a meaning for us, it is because they derive it from our experience as embodied subjects." (Merleau-Ponty, 1995: 182), or speak of a corporeal dwelling: "Our body is intimately tied to what we walk on, sit on, touch, taste, smell, see, breathe, and move within. Our corporeality is part of the corporeality of the world" (Lakoff & Johnson, 1999: 503). Understanding it presupposes a knowledge of the incarnate subject that we are knowing the absolute existence of the object as an alternative to the thought of an absolute object, that in the reflexive analysis. The absolute existence of the object is the way I dwell: constituting the object, in other words, I constituting it like – by the same principle – my body, it is already given to me, without systematically seeing it, I do not have it to think of him, he exists in me: "If there is, for me, a cube with six equal sides, and if I can link up with the object, this is not because I constitute it from the inside: it is because I delve into the thickness of the world by perceptual experience." (Merleau-Ponty, 1995: 182).

5 CORPOREAL VALUE

Merleau-Ponty, starting from the phenomenology of the body initiated by Husserl, reaches the value of the corporeal instead of the symbolic, and we may identify it as the evolutionary value of dwell in experience: "an I not in the sense of an absolute subjectivity, but indivisibly demolished and remade by the course of time." (Merleau-Ponty, 1995: 196), in counterpoint with the crystallized value of dwell in permanence. The value of the corporeal is not an absolute value – immutable – it is drawn in every life, in every body, in every gesture, and is born every day, at all times, with the new being. It does not establish itself with the permanence of the building (Rossi, 1966: 49), but is experienced, in each new act of seeing, intuitively, as in an 'architecture of action' (Montaner, 2017: 131):

> "During the performance, the notes are not only the 'signs' of the sonata, but it is there through them, it enters into them. In the same way the actress becomes invisible, and it is Phaedra who appears. The meaning swallows up the signs, and Phaedra has so completely taken possession of Berma that her passion as Phaedra appears the apotheosis of ease and naturalness." (Merleau-Ponty, 1995: 164).

When Figueiredo says: "the building was like a construction that already existed there" (Figueiredo in Arenga, 2012: 48), the architect doesn't only mention the symbolic value of the site, according to theories of the architect Christian Norberg-Schulz (1926–2000) – the spirit of the place – understood as being (subject-author or subject-client) – a perceptual experience –, and therefore, as Figueiredo states, it exists independently of the program that is defined, in other words, is a piece of that world that I experiment with Mitra [in my body].

I do not idealize an experience. That nature (essence) that is symbolized as a memory of a reality (life), is, alternatively, embodied in Figueiredo's architecture. It is experience and not symbol. An image – we may affirm –, but manufactured in the experience, as the architect's expression: "...listening to music, for me, is something with a certain volume and it's like reading a book or watching a movie: it's an act, I'm enjoying it." (Figueiredo in Arenga, 2012: 46)

The value of the corporeal is the encounter with the true Cogito, the existence of which is not thought of its existence, but rather the revelation of the being-in-the-world, or in other words, his experience as contact with the world. The value of the corporeal exists, if between thought and thing – between idea and object – there is no separation, but an articulation, a communication, through my body as expression of that: it [body] touches both simultaneously. Mitra speaks in me:

> "To perceive is [...] to see, standing forth from a cluster of data, an immanent significance..." (Merleau-Ponty, 1995: 19); "...in the full sense of the word (as the antithesis of imagining) is not to judge, it is to apprehend an immanent sense in the sensible before judgement begins." (Merleau-Ponty, 1995: 31).

REFERENCES

Arenga, N. 2012. Vitor Figueiredo: Fragmentos de um Discurso. Circo de Ideias.

Bollnow, O.F. [1963] 2011. Human Space. Translated by Christine Shuttleworth and edited by Joseph Kohlmaier. Hyphen Press – London.

Husserl, E. [1906/07] 2008. Introduction to Logic and Theory of Knowledge. Lectures 1906/07. Translated by Claire Ortiz Hill. Springer.

Husserl, E. [1929] 1975. The Paris Lectures. Translated by Peter Koestenbaum, Kluwer Academic Publishers.

Husserl, E. [1902/03] 1999. The Idea of Phenomenology. A translation of Die Idee der Phanomenologie Husserliana II by Lee Hardy, Kluwer Academic Publishers.

Ingold, T. 2000. The Perception of the Environment_ Essays on Livelihood, Dwelling and Skill. London, Routledge.
Lakoff G. & Johnson, M. 1999. Philosophy in the Flesh. The Embodied Mind and its Challenge to Western Thought. Basic Books.
Lyotard, J-F. 1991. Phenomenology. Albany: State University of New York Press.
Merleau-Ponty, M, [1945] 2005. Phenomenology of Perception. Translated from the French by Colin Smith. Routledge.
Minkowski, E. [1933] 1970. Lived Time. Phenomenological and Psychopathological Studies. Translated by Nancy Metzel. Northwestern University Press.
Montaner, M. 2017. Do diagrama às experiências, rumo a uma arquitetura de acção. Barcelona, Gustavo Gilli, SL.
Nesbitt, K. 2006. Uma Nova Agenda para a Arquitectura. Antologia Teórica (1965–1995). Tradução de Vera Pereira. São Paulo, Cosac Naify.
Pallasmaa, J. [1996] 2011. Os Olhos da Pele. A arquitectura e os sentidos. Bookman.
Pallasmaa, J. 2013. As Mãos Inteligentes. A sabedoria existencial e corporalizada na arquitectura. Bookman.

Drawing (...) City (...) Body, Dwelling on Earth – Janeiro (Ed.)
 ISBN 978-0-367-02602-8

(Para)metric representations and urban indicators: An application of citymetrics in a neighborhood of João Pessoa, Brazil

F. Lima
Department of Design, Representation and Technology, Faculty of Architecture and Urbanism, Federal University of Juiz de Fora, Juiz de Fora, Brazil

G. Silva
Department of Architecture and Urbanism, Center of Technology, Federal University of Paraíba, João Pessoa, Brazil

ABSTRACT: This article presents a computational approach to support decision making in urban design tasks, in order to provide dynamic assessment and optimization of indicators related to principles derived from Transit Oriented Development (TOD) – an urban development model that combines regional planning, city revitalization, suburban renewal, and walkable neighborhoods. Nevertheless, in spite of being increasingly promoted in several cities of the world, TOD lacks an approach that addresses multivariable data for optimization of its principles. In this paper, we implement a computational system entitled CityMetrics in a neighborhood unit in a case study. The objective is to verify the possibilities of improvement of urban configurations performance by addressing TOD relevant and measurable features (transit accessibility, walkability and diversity) in a (para)metric representation, evaluation and optimization logic. The ultimate goal is to facilitate the management of solutions in urban planning processes, seeking to provide dynamic management of intervention possibilities.

Keywords: Parametric Urban Design, Transit Oriented Development, Sustainable cities

1 INTRODUCTION

Motor vehicles fuel combustion is responsible for up to 75 percent of urban air pollution (Global Fuel Economy Initiative, 2016), while outdoor air pollution was associated with 3.7 million premature deaths in 2012 (World Health Organization, 2014). In this context, transportation is one of the largest carbon emitters on earth.

Besides contributing to global warming, the adopted organization logic of cities is responsible for some inconveniences in contemporary urban life, such as automobile dependence, traffic jams and less social interactions. In this regard, Gotschi and Mills (2008) state that walking and bicycling for the shortest trips (less than 1 mile) rather than taking a car, could reduce carbon dioxide emissions by 12–22 million tons per year, only in the US.

In this scenario, Transit Oriented Development (TOD) is an urban development model that advocates a paradigm shift in the logic of dwelling on cities, as long as TOD advocates the creation of compact, walkable, mixed-use neighborhoods, centered around high-quality transport systems. That is, TOD makes possible the creation of vibrant livable communities without depending on automobiles for urban mobility, and it is being increasingly promoted in several cities of the world as a sustainable policy (Calthorpe, 1993; Vale, 2015).

Actually, there is not a universally accepted definition of TOD, although it is frequently described in a physical way: a compact mixed-use buildings area, within walkable distance to a multimodal mobility network (Cervero & Kockelman, 1997; Calthorpe & Fulton, 2001; Suzuki et al., 2013; Vale, 2015). Therefore, in TOD's logic, basic urban needs are easily accessible, with no need to demand automobiles or to spend large amounts of time in commuting.

TOD can be characterized by a few main features and its primary (and measurable) principles consist of: (a) **transit accessibility** – locating amenities, employment, retail shops and housing around transit hubs; (b) **walkability** – connecting houses and amenities through distances that can be traveled on foot or by bicycle; (c) **diversity** – providing a mix of uses, housing types and densities in the same neighborhood; (d) **density** – stimulating infill and redevelopment along transit corridors within existing neighborhoods, allowing transportation systems to run efficiently (Cervero, & Kockelman, 1997; Calthorpe & Fulton, 2001; Dittmar & Ohland, 2004; Suzuki et al., 2013).

In this context, TOD defends proximity to transport stations and a functional relationship with them, as well as mixed-use and compact neighborhoods that stimulates walking, cycling, and the use of public transportation by residents, employees, shoppers and visitors. In other words, TOD seeks to boost the use public transportation, by creating urban areas that are compact, diverse and has an urban design geared to the pedestrian – where one can walk to the stations and other urban needs.

However, and according to Dittmar et al. (2004), there have not been sufficiently developed approaches or systems to help the involved actors in the process of providing successful TOD projects. Nevertheless, TOD is a multivariable dependent model, as long as it relies on geometric and measurable metrics for improved performance. It considers geometric principles and measurable parameters for designing autonomous neighborhoods. In other words, it represents a kind of complex proposition, derived from some variables considered as crucial in achieving its objectives, what makes TOD a potential case for computational implementation (Lima et al., 2016).

Architectural (and urbanistic) practice, in turn, has been faced with a paradigm shift towards the application of digital methods in design activities. In this regard, it is a challenge to promote a suitable conceptual and instrumental background for making our cities better, using resources, tools, representations and instruments to provide more sustainable ways of dwelling on earth.

Thus, the scope of this paper refers to a context of intersection between computational representations – more specifically those that concern algorithmic and parametric logic – and the implementation of measurable attributes (or metrics) related to a more sustainable performance of urban configurations.

In this context, this paper presents the application of CityMetrics (LIMA, 2017), a computational system that considers TOD related metrics for support in urban planning. This system was used to run analysis of a Brazilian neighborhood, in the city of João Pessoa, Paraíba.

Accordingly, this paper is structured in the following sequence: (a) a brief description and presentation of CityMetrics, and its set of tools; (b) an implementation (case study) of the proposed system in the Altiplano neighborhood, in João Pessoa; (c) a discussion about the obtained results, and; (d) an overall discussion and final remarks.

2 CITYMETRICS

CityMetrics (LIMA, 2017) is intended to assist on TOD oriented planning processes and is not wanted to act as an independent-automatic solver. In this sense, the role of the many players involved in urban planning tasks remains central, and it's still these players that will stablish objectives, feed the system and consider "non-programmable" and subjective aspects.

Besides, CityMetrics is developed to measure and optimize urban configurations performance through metrics related to TOD principles, and consequently, to help in proposing more sustainable cities. In this sense, the proposed system consists of the following developed tools: (i) Physical Proximity Calculator – measures the Physical Proximity Index (PPI), an indicator that calculates the smallest physical distances (considering slopes) between the nearest target (station or amenity) and one (or all) origin(s) in a neighborhood; (ii) Topological Proximity Calculator – contemplates Space Syntax principles for calculating the smaller topological distances (number of steps) between a target (station or amenity) and one (or all) origin(s) in a neighborhood; (iii) Amenities Variety Calculator – performs the calculation of the average physical distances (considering slopes) between all targets (amenities) and one (or all) origin(s) in a neighborhood; (iv) Amenities Recurrence Calculator – calculates the ratio of the number of targets (amenities) and the total number of origins of an examined area; (v) Mixed-use index calculator – computes the ratio between residential and non-residential areas in a location, in order to analyze its diversity; (vi) Spacematrix calculator – calculates Spacematrix density indicators in order to support goals definitions and decision-making processes.

2.1 *Physical Proximity Calculator (PPC)*

Physical Proximity Calculator (PPC) is an algorithmic tool that measures the distance between a target (transport stations and/or amenities) and one (or all) locations(s) in a neighborhood (origins). In this regard, the proposed algorithm calculates the path(s) with smaller physical distance(s) between a target and one (or all) destination(s) in a district, considering slope(s) in the path(s). Thus, if a given plot (origin) is within 400 m (5 min walk) from a target, it is assigned with a value of 1. The score decreases as the distance approaches 1.6 km (20 min walk) and a 0 index is awarded for distances greater than 1.6 km. PPC can be used to measure physical proximity of many amenities categories, but in this work, will be considered: food, retail, educational, entertainment and recreation. The average score from all these categories gives a final index of Physical Proximity.

2.2 *Topological Proximity Calculator (TPC)*

TPC is a tool that calculates proximity considering topological metrics, using concepts from Space Syntax theory (Hillier & Hanson, 1984). This tool can be used in a complementary way to the tools that use the physical metrics. In this sense, the TPC aims to calculate: (a) the pathways with the smallest topological distances between origins and targets of a locality, and; (b) the integration/depth of the spaces of a given area. That is, this tool calculates the number of direction changes (topological steps) needed to reach one or more targets from a given origin, and also indicates which spaces are more integrated – which means identifying which streets are more accessible and, consequently, have greater relevance in the dynamics of an urban area.

2.3 *Amenities Variety Calculator (AVC)*

The AVC measures another important aspect for the evaluation of the walkability of a given location: the proximity between an origin and all the targets (amenities) within walkable reach. In this sense, this tool calculates the average distances between a given source and all the nearby targets in a given category of urban services. AVC algorithm works in a complementary way to PPC because, while the former considers only the distance to the nearest amenity, the AVC assigns an index considering the average physical distance between all the targets and the origin in question. Thus, while PPC measures the distance from the closest service to an origin, AVC considers the distances between the same source and all the targets reported in the same category.

2.4 *Amenities Recurrence Calculator (ARC)*

Amenities Recurrence Calculator (ARC) in another developed tool that aims to measure the walkability of an area and works in a complementary way to PPC and AVC tools. While the formers calculate the lowest physical distance to the nearest target and the average physical distance to all targets, respectively, ARC has quantitative characteristics, as it calculates the proportion of the number of targets reported (in each category of services) and the total number of locations in a surveyed area.

This algorithm, which basic operation consists in counting targets within a radius of 20 minutes of walking and dividing this value by the number of lots in the same area, is important for analyzing the supply of services in each neighborhood, an aspect equally important to measure the capacity that a given locality has to connect dwellings and various urban services by distances that can be traveled on foot.

2.5 *Mixed-Use Index Calculator (MXIC)*

The MXIC aims to measure the diversity of a neighborhood. For this, MXIC incorporates the mixed-use index (MXI) concept (Hoek, 2008) that calculates the proportion between the sum of all the residential and non-residential areas of a locality, making a comparison of these proportions. The closer the ratio between areas is 50/50, the greater the diversity an urban area has. This algorithm has great importance in the context of CityMetrics, because it allows one to assess the diversity of a certain area before and after potential interventions.

2.6 *Spacematrix Calculator (SPC)*

The Spacematrix calculator algorithm implements codes for calculating density attributes from studied areas, informing three fundamental indicators proposed by Pont & Haupt (2010): Intensity – Floor Space Index (FSI); Coverage – Ground Space Index (GSI) and Network Density (N). Thus, once the required inputs are informed (contours of the analyzed area and its buildings, number of buildings floors, and design of the streets) the algorithm informs the FSI, GSI and N of an analyzed area. This tool has great importance in the context of CityMetrics, since it allows to calculate and visualize information about the density of urban areas in real time, allowing a dynamic evaluation of urban contexts studied and the proposed interventions/modifications.

3 THE CASE STUDY

The case study comprises the application of the CityMetrics set of tools on the existing district of Altiplano, in the city of João Pessoa, Paraíba state, Brazil (see Figure 1). These studies were developed during a course, taught by the authors of this article, in the Architecture and Urbanism postgraduate program from Federal University of Paraíba (PPGAU-UFPB). The main goal of these studies was to verify the possibilities of improvement of the addressed area, considering the scope of the aforementioned TOD principles.

Figure 1. Aerial views of Altiplano district.

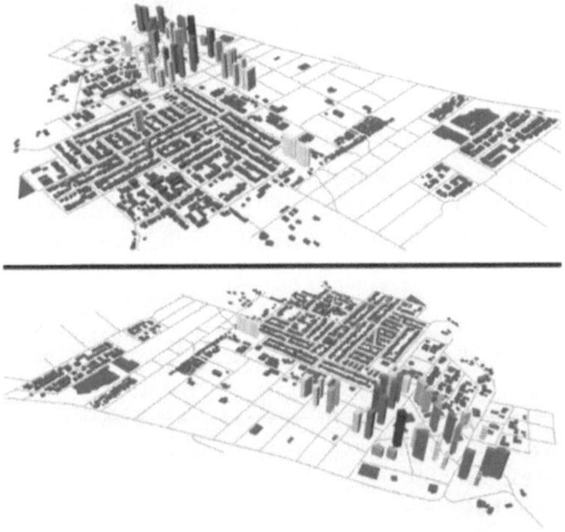

Figure 2. The parametric model of the district, in Grasshopper/Rhinoceros software: the network of streets, buildings, available areas, density indicators, uses and amenities positions that can be assessed in a dynamic way.

The Altiplano district is a predominantly residential neighborhood and presents the following features: (a) a suitable extension for TOD implementation (approx. 1 km diameter); (b) a relatively low density – high verticalization in some plots but no balanced distribution; (c) no transport station (only a few bus stops and only two lines available in the whole district); (d) many available plots for new buildings and developments; (e) a very unbalanced relation between residential and non-residential areas. In summary the Altiplano neighborhood does not meet the advocated logic by TOD principles.

In this context, the following information of Altiplano was obtained and transferred to a digital-parametric analysis model: (a) the footprints, number of levels, uses (residential and non-residential) and topographical positioning of each building inside the district, in order to measure distances, slopes, possible connecting paths, diversity and density indicators; (b) the location of each one of the amenities on the neighborhood – according to the aforementioned categories – in order to measure proximity, diversity and variety of urban services; (c) the identification of available areas for new buildings (vacant lots and non-consolidated places), seeking to provide room for new constructions and to modify the diversity of the district; (d) The topographical network of neighborhood's streets, with the view to consider slopes and distances for measuring PPC, AVC and ARC operations; and (e) the design of streets, blocks and lots, in order to provide density (Spacematrix) assessment. Figure 2 illustrates the developed model.

Thenceforth, during the classes, the students were introduced to CityMetrics logic and were organized into four groups that should use the presented tools to address the studied area. The task was to apply simulation and optimization operations, aiming at an increase of transit accessibility, walkability and diversity related indexes, by the following actions, respectively: (a) **transit accessibility** – look for the best location for insertion of new bus stops, in order to increase Physical Proximity indexes between all plots in the neighborhood and the nearest bus stop available; (b) **walkability** – insert only one amenity from each category, seeking to increase global PPC, AVC and ARC with the lowest number of new services possible; (c) **diversity and density** – evaluate different occupation strategies for vacant lots, considering Mixed-use index and Spacematrix indicators to help analyzing impacts from new buildings propositions.

4 RESULTS

The implementation of CityMetrics provided some analysis and new arrangements for the neighborhood, with better performance from the scope of TOD principles. In summary, the groups analysis and propositions provided: (a) a configuration that suggests a better **transit accessibility**, since the optimized insertion of a bus stops permitted a higher average Physical Proximity score to them; (b) an arrangement that suggests a greater **walkability**, as long as the optimized addition of new amenities provided an increase of the neighborhoods proximity, diversity and variety, in all analyzed categories; (c) various scenarios for a neighborhood with more **diversity**, since the proposition of new buildings and their functions provided a more balanced Mixed-use index, seeking a greater equilibrium between residential and non-residential places; (d) different contexts for **density** evaluation using Spacematrix indicators,

Table 1. Results obtained before and after CityMetrics implementation.

	Before	After
Average Physical Proximity to the nearest bus Stop – **transit accessibility**	0.93	0.99
Average Physical Proximity to the nearest amenity (food) – **walkability**	0.85	0.90
Average Physical Proximity to the nearest amenity (entertainment) – **walkability**	0.56	0.92
Mixed-use Index (resid/n-resid) – **diversity**	0,97/0,03	0,51/0,49

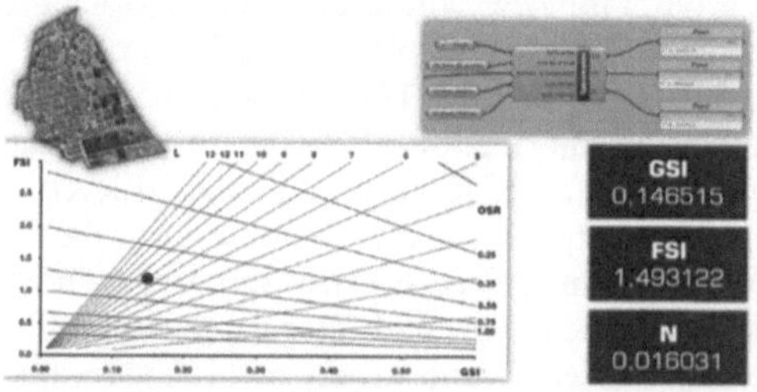

Figure 3. Spacematrix indicators of the neighborhood in its current configuration. The SPC tool allowed a dynamic evaluation for different propositions of new buildings.

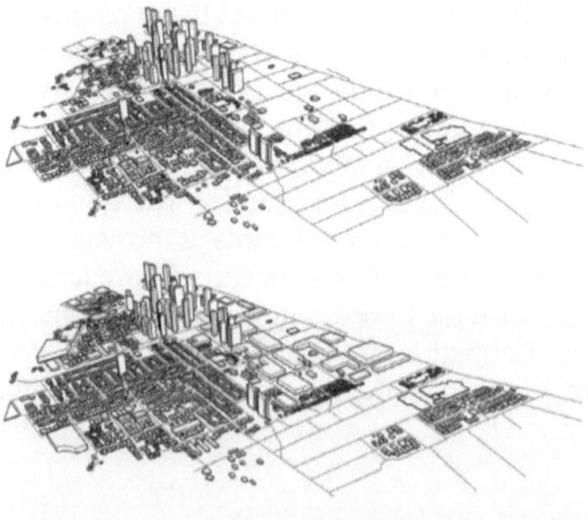

Figure 4. The original configuration of Altiplano neighborhood (above) and a new occupation scenario, proposed in the discipline, after CityMetrics implementation (below): better indicators of transit accessibility, walkability and diversity.

since parametrically controlled urban geometries made possible to regulate density, in a manner that keeps more people (working or living) closer to transport hubs and to other urban services. The following table and figures illustrate the obtained results:

5 DISCUSSION AND FINAL REMARKS

Citymetrics proved to be helpful in order to provide dynamic measurement, assessment and optimization of principles derived from TOD. The positioning of the new bus stops and amenities was supported by optimization tasks that allowed to identify, among hundreds of options, the solutions that provided the lowest average distance between these urban services and all lots in the neighborhood. In this sense, the obtained PPC index was as high as it could get, considering the district's configuration and the limitation of proposing only one new amenity for each category. Besides, the locations with worse PPC scores were easily identifiable, and in this sense, this is a good starting point for further actions for transit accessibility improvement (the adoption of secondary modals for connecting these locations and the station, for example).

Despite of not being directly optimized, diversity and density indicators played a fundamental role in the elaborated studies, as long as they allowed visualizing and evaluating different scenarios for vacant plots occupation, guiding interventions and giving hints from building potential and mix of uses distribution. The MXC and SPC indicators proved to be useful in the developed implementation, given that they consider objective and measurable features for assessment of diversity and density, respectively.

5.1 *Limitations, further developments and final remarks*

Despite the usefulness of a system like CityMetrics for supporting TOD oriented urban planning processes, we recognize some limitations in the presented approach. Firstly, we admit the the addressed indexes does not fully incorporate the diverse features that can influence the walkability of an urban area, for an example. Secondly, density has a fundamental role within the scope of TOD. In this context, variables related to populational density and the implementation of modals in a neighborhood should also be tested. Thirdly, it is very important to consider the limitations of the presented approaches, especially if we consider that it was elaborated throughout a discipline. However, the present work points to some interesting perspectives of dynamic simulation that computational representation brings to urban design possibilities.

This work seeks to contribute to a better understanding of the benefits and drawbacks of handling (para)metric representations and complex data relationship in objective TOD planning tasks, in order to outline a starting point for computational approaches towards more sustainable ways of planning cities and dwelling on earth. In this sense, this work aims to demonstrate the approach's potential towards a more efficient TOD implementation methodology.

ACKNOWLEDGEMENTS

The authors would like to thank the students involved in this study and the Faculty of Architecture and Urbanism of Federal University of Juiz de Fora, for financial support for presentation of this work.

REFERENCES

Calthorpe, P. 1993. The Next American Metropolis: Ecology, Community and the American Dream. New York: Princeton Architectural Press. New York.

Calthorpe, P. & Fulton, W. 2001. The regional city: planning for the end of sprawl. Washington, DC: Island Press.

Cervero, R. & Kockelman, K. 1997. 'Travel demand and the 3Ds: density, diversity, and design', Transportation Research Part D: Transport and Environment, 2, pp. 199–219.
Dittmar, H. & Ohland, G. (eds.) 2004. The new transit town: Best practices in transit-oriented development. Washington, DC: Island Press.
Global Fuel Initiative. 2016. Available in: http://www.globalfueleconomy.org/.
Gotschi, T. & Mills, K. 2008. Active Transportation for America: The Case for Increased Federal Investment in Bicycling and Walking. Railsto-Trails Conservancy, Washington DC, 44 pp.
Hillier, B. & Hanson, J. 1984. The social logic of space. Cambridge: Cambridge University Press.
Hoek, J.V.D. (2008) 'The MXI (Mixed use Index). An instrument for anti-sprawl policy?', Proceedings of the 44th ISOCARP congress 2008.
Lima, F.T. 2017. Métricas Urbanas: sistema (para)métrico para análise e otimização de configurações urbanas de acordo com métricas de desempenho. PhD Thesis: Federal University of Rio de Janeiro.
Lima, F.T., Kós, J.R. & Paraízo, R.C. 2016. Algorithmic approach toward Transit-Oriented Development neighborhoods: (Para)metric tools for evaluating and proposing rapid transit-based districts', International Journal of Architectural Computing, 14(2), pp. 131–146.
Pont, M.B. & Haupt, P. 2010. Spacematrix: Space, Density and Urban Form. Rotterdam: NAi Publishers.
Suzuki, H., Cervero, R. & Iuchi, K. 2013. Transforming cities with transit: Transit and land-use integration for sustainable urban development. Washington, DC: Word Bank.
Vale, D.S. 2015. 'Transit-oriented development, integration of land use and transport, and pedestrian accessibility: Combining node-place model with pedestrian shed ratio to evaluate and classify station areas in Lisbon', Journal of Transport Geography, 45, pp. 70–80.
World Health Organization. 2014. 7 million premature deaths annually linked to air pollution. Available in: http://www.who.int/mediacentre/news/releases/2014/air-pollution/en/.

Drawing (...) City (...) Body, Dwelling on Earth – Janeiro (Ed.)
© 2019 Taylor & Francis Group, London, ISBN 978-0-367-02602-8

Hundertwasser five skins: Pretext for the study of the role of drawing and human body in the project of architecture

F. Moreira da Silva
CIAUD, Lisbon School of Architecture, Universidade Lisboa, Lisbon, Portugal

ABSTRACT: The study by drawing uses several means and processes of abstraction, through the modeling by the line and by the stain, manipulation and production of the image, providing a critical analysis of the information obtained in the sense of understanding the human body, central element in the architecture project. The scale of the body from the perspective of the Austrian artist and architect Friedensreich Hundertwasser is the starting point for this investigation about the role of Drawing on the understanding of the Human Being and from this one with the world around him. We pretend to underline the learning process through Drawing, underlining the relevance, vitality and permanence of this way of thinking and acting in the architects' training.

Particular attention is given to the typology of the line and the structuring and appropriation of the form. The gesture, also seen in the sketching as a process of capturing the essence, in an attitude of critical analysis, of understanding, has within it unique potentialities, such as its operative capacities of acquisition and communication of information.

With this research project we intend to develop a reflection on the human body, its dimensions, forces, biomechanics, using the use of various technical and graphic expressions. Undoubtedly, drawing as a visual and organizational means of communication also stimulates the user as a tool and as an irreplaceable means of personal expression. Drawing becomes thus a language and a support of critical analysis of the human body.

Keywords: Drawing, Human Body, Hundertwasser, Architecture

1 INTRODUCTION

The present research project presents a reflection on the human body, its dimensions, forces, biomechanics, using the use of various technical and graphic expressions. Undoubtedly, drawing as a visual and organizational means of communication also stimulates the user as a tool and as an irreplaceable means of personal expression. Drawing becomes thus a language and a support of critical analysis of the human body.

Using the vision of the painter and architect Undertwasser about the five skins as pretext, we want to reflect about the role of drawing and the human body in the project of architecture.

2 HUNDERTWASSER FIVE SKINS

For Hundertwasser, the Human Being has five skins: the epidermis, the clothes, his house, the identity and the Earth.

When, in 1967 and 1968, the artist delivered his "Naked" speech to proclaim the right of the Human Being to his third skin (the free alteration of his home), he performed the complete ritual cycle of his spiral. He rediscovered his first skin that of his original truth, his nakedness as a man and a painter, taking his second skin (his clothes) to proclaim the right to his third skin (his house).

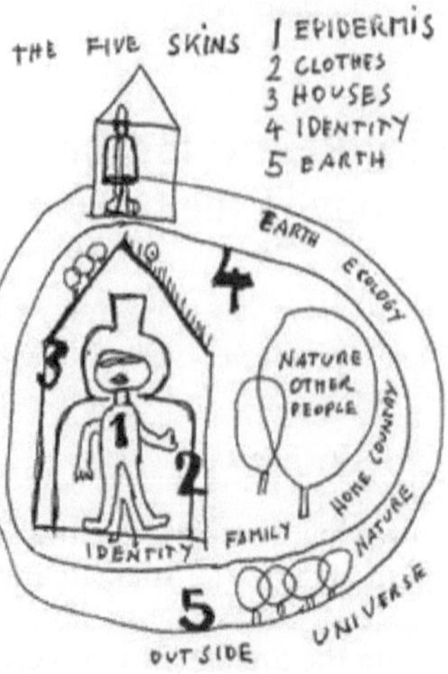

Figure 1. Hundertwasser Five Skins in http://www.hundertwasser.com/skin, accessed February 5, 2018.

Later, after 1972, his consciousness of the Human Being was enriched by new questions, which demanded new answers and made new commitments. Then it came the new covers that should be added to the concentric wrapping of the previous three. The fourth skin of the Human Being is the social environment (of family and nation, through the elective affinities of friendship). The fifth skin, which is the planetary skin, is directly concerned with the fate of the biosphere, the quality of the air we breathe and the state of the earth's crust that shelters and feeds us.

This approach of the Austrian artist and architect Friedensreich Hundertwasser was the starting point for this investigation about the role of Drawing on the understanding of the Human Being and from this one with the world around him.

3 DRAWING AND KNOWING TO DRAW

> *"The drawing prolongs the act of the hand, and with it the wrist, the forearm, the gaze, and finally the whole body. Against the intelectulalisation, which has sometimes been reduced, the drawing produces a rhythmic configuration of reality arising from the rhythms of the body itself."* (Chalumeau, 2008: 8)

There is a difference that needs to be clarified, which is between Drawing and Knowing to Draw. Knowing to Draw is more focused on the truth of error than on the truth of certainty. In this sense, to know how to draw is to live with the error by integrating it, taking into account the expressive potentialities and the truth that it contains, a truth that is based on creative identity. (Pedro, 2012)

Besides supporting us on the critical sense and the discourse, drawing is not based on free will, but on a solid basis of formation, using methodological processes adapted to different needs, without the stigma of fear and prejudice.

Attention must be paid to meaningful content, to an intervention at the level of analysis and the logical validity of representation. Drawing goes far beyond the exploration of material techniques, the contingencies of gesture and inner vision. Knowing how to draw is not exhausted in the emotional component, in the so-called emotive gesture, since different cultural states correspond to different emotional states that is, there is a causal relation between culture and emotion, and there is an intimate relationship between emotion and knowledge. (Hall, 1966)

The Drawing that we approach has more to do with the Knowing how to Draw, with the fact that it is a privileged aesthetic-visual means of apprehending the world, of what surrounds us, of understanding the Human Being and its habitat, the architectural space.

Thus, there are several variables to meet when it comes to know how to draw, such as intelligence, culture, technical knowledge, aesthetic sensibility, talent, experience. Drawing, in its sense of drawing, does not depend on a gift, but on learning in the field of visual literacy, allowing the exploration of new forms of expression, which are not those of reproducing external reality.

Drawing is a factor in the development of creativity and imagination, not being held hostage by currents and artistic movements, in a deep cut with what was an iconographic tradition that served as its base, but in the language of other periods one can find strong motivation for the discovery of new forms, means and processes of registration. (Pedro, 2015)

4 DRAWING AND HUMAN BODY

> *"Drawing is a magical spirit that is everywhere and nowhere, its existence by the line is unknown in nature, however it is able to suggest us volumes and grandiose spaces in small spaces: fixed forms and their composition the side elected more by history than by the sculptor and which the observer hardly forgets."* (Gastão, 1989: 39–40)

Drawing allows the development of reading, understanding, research, reflection and formal appropriation capacities, through the use of graphic expressions characteristic of the sketch, as well as of communication, being an essential instrument in the conception and representation in architecture: thinking and communicating through the design ideas underlying the project.

The drawing of the human body, in the whole or in the parts that compose it, can be a means of approximation and simulation to the architecture.

That's why Hundertwasser established a strong relationship between the body and the habitat. Architecture must fulfill all the requirements inherent to the use that the Human Being intends to give to a certain space, being a vehicle for happiness.

We must understand the interactions between human beings and the elements of a system, which currently is meant for all aspects of the human activity.

Figure 2. Pedro Janeiro's Sketch during a Drawing course, when student, author's data collection.

So, Drawing as a privileged instrument of thinking, is an indispensable tool in Architecture and in the whole conceptual process attached to the project itself.

This way there are two fields of knowledge which immediately cross themselves, autonomously: drawing, viewed as idea, as a pre-figuration, such as sketch; design, in its broadest sense of project, as a descendant of drawing.

It's important to highlight the vitality of Drawing as a means of approaching when projecting, reason why architects must know and be able to manipulate it.

The Human Figure Drawing is a way of understanding the Human Being and his/her relationship to what surrounds him/her.

Drawing allows the development of reading, understanding, research, reflection and formal ownership skills, through the use of graphical expressions of the own sketch drawing, as well as of communication, being an essential tool in the conceptual and representation phases in architecture.

Besides being a form of registration, Drawing is also a way of organizing the thought, having an interdisciplinary character, having a primordial role in all the phenomena of representation and description. It is a gesture, a drive that leaves a trace, a mark in space, imprinting its temporal passage: the gesture in the Drawing is a determined action and launched at a time of its own.

At the service of an intention, the energy of the action justifies the indicia dimension inherent to the track. Derrida (2010: 53) refers to the still not visible part of the trait, the blindness of the trait, at the time of its original disruption, in the tracing power of the trace, both in what is traced and is track, as in what has not yet been drawn, given that in Drawing we also play with what has not yet been drawn, which is not yet visible, and what still is not defined by the already drawn.

In architecture, Drawing works as a basic element of the project, as a basic support and as an element of aesthetic coherence. Given that the Human Being is the measure of all things, in architecture the representation of the human figure is central, starting from the model in favour of mimesis.

Through Drawing it is possible to construct the sense of place that implies the human scale, in its multiple dimensions.

The consolidation of design as an autonomous discipline guided by a theoretical and technical body is relatively recent, especially in relation to disciplinary discussion.

To draw is to act from the visible and the experienced, or from a combination of terms. If memory is important as an ability to retain, retrieve, store and evoke information, it is also important to experience the present-continuous, implicit in works developed with real space. (Gantes, 2012)

Drawing grammar gives particular attention to line typology and the structure and appropriation of the form.

The gesture, the movement, also seen in the sketch drawing as an attitude of critical analysis, of understanding, contains in itself some unique capabilities, such as its operational capabilities for acquisition and communication of information.

Figure 3. Student's Drawing, author's data collection.

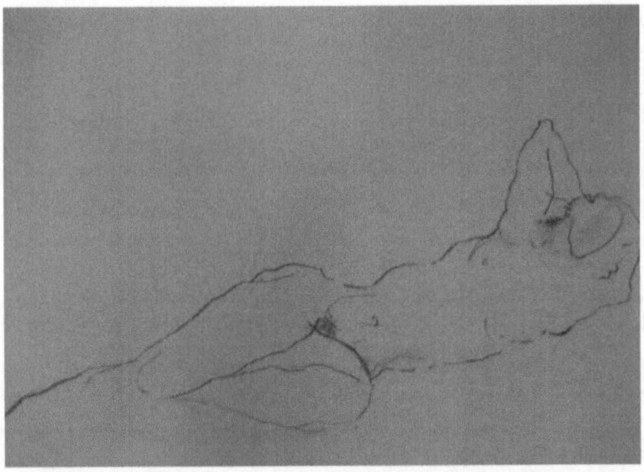

Figure 4. Pedro Janeiro's Sketch during a Drawing course, when student, author's data collection.

With the development of spatial vision, we are able to analyse, understand and represent synthetically what's around using the same sketch process for the project-oriented ideation in architecture.

The study by drawing uses several means and processes for capturing, by modelling through line and by staining, image manipulation and production, providing a critical analysis of the obtained information in order to understand the human body, central element in architecture project.

This way one can get an approach to architecture project through simulations and specific practices which aim mainly the reflection on the human body, its dimensions, strength, biomechanics, resorting to the use of different technical and graphical expressions.

There's no doubt that Drawing as a means of visual communication and organizational concept also encourages the user as tool and irreplaceable means of personal expression.

This way, Drawing becomes a language and a support of critical analysis of the human body, and central for the process of architecture project.

5 CONCLUSIONS

Drawing as an act, drawing is an intense way of living time, an absolutely current time, not measurable by the clock.

The study by drawing uses several means and processes of abstraction, through the modeling by the line and by the stain, manipulation and production of the image, providing a critical analysis of the information obtained in the sense of understanding the human body, central element in the architecture project. The scale of the body from the perspective of the Austrian artist and architect Friedensreich Hundertwasser was the starting point for this investigation about the role of Drawing on the understanding of the Human Being and from this one with the world around him.

His perspective of the world was essential to place the Human Being in the centre of the process, showing that in architecture we have not only to deal with Human Beings, but that we project for them as final users of the projected spaces.

We also underlined the learning process through Drawing, showing the relevance, vitality and permanence of this way of thinking and acting in the architects' training, giving particular attention to the typology of the line and the structuring and appropriation of the form. The gesture, also seen in the sketching as a process of capturing the essence, in an attitude of critical analysis, of understanding, has within it unique potentialities, such as its operative capacities of acquisition and communication of information. The sketches of the Human Body are much more than Drawings, because they serve as a fundamental instrument of reflection on the interaction of Human Being/Space, which is central for the project of architecture.

Drawing is the interpretation of the visa through the gesture that condenses, or simplifies, the time of the experience. It is the quintessential way of thinking visually.

Its importance manifests itself not only as an end in itself, but also as a means of exercising the capacity to think what we do and the relationship with the world itself.

REFERENCES

Chalumeau, J.-L. 2008. Les 200 plus beaux dessins du monde, Paris: Hachette.

Derrida, J. 2010. Memory of the Blind, Self-Portrait and Other Ruins, Lisbon: Calouste Gulbenkian Foundation.

Gantes, M. 2012. O Desenho em Aberto, In: Desenhar, saber desenhar, pp. 124–129, Lisboa: Faculdade de Belas Artes.

Hall, E.T. 1966. The Hidden Dimension, NY: Doubleday/Anchor.

Hundertwasser, Friedensreich, Five Skins in http://www.hundertwasser.com/skin, accessed February 5, 2018.

Pedro, A. 2012. Desenho: (In)definições e (In)certezas, In: Desenhar, saber desenhar, pp. 96–107, Lisboa: Faculdade de Belas Artes.

Pedro, A. 2015. A expressão do tempo no desenho, In: As idades do desenho, pp. 113–122, Lisboa: Faculdade de Belas artes, CIEBA.

Marques, G. 1989. Encontros com António Duarte. Lisboa: Imprensa Nacional—Casa da Moeda.

Drawing (...) City (...) Body, Dwelling on Earth – Janeiro (Ed.)

The mask of the classical theater for a Tactile Museum, through artisanal approaches and digital protocols

F. Fatta & A. Marraffa
Department Architettura and Territorio (d'ArTe), Mediterranea University of Reggio Calabria, Italy

ABSTRACT: The mask is a magical tool, rich of values, meanings and functions. The ritual function prevails above all, because it represents a device able to introduce the person who wears it inside a mystic and extraordinary world. A huge collection of terracotta artifacts is stored in the 'L. Bernabò Brea' Archeological Museum of Lipari. It represents the widest and completest corpus about the theatrical mask and about the scenic costumes, ever arrived from the Classic antiquity, a real 'material deposit for the theatrical memory'. Thanks to those founds it is possible to retrace almost two centuries of Greek theater: from the Age of the great tragedies of Sophocles and Euripides, until the Menander's New Comedy. During a three-years Research, we tried to appreciate the historical profoundness that foments the material culture of theater, the tradition of a civilization that from the V century has been linked to the masks, first of all for a theatrical use, then for a ritual one, as funeral supply. This second use allowed to preserve the artifacts, revealing all the iconographic features of the Tragedy and the Comedy's characters, belonging to the Classic world. Thanks to an advanced photogrammetric survey and to some new digital techniques, we analyzed many mask's fragments with the aim of the knowledge, of the documentation and of the dissemination of the material culture of theater. We studied almost a thousand of fragments and we tried to draw up an useful methodology to rebuilt and to reintegrate, in a digital way, the terracotta founds. The digital workflow, of reconstruction and anastylosis, was also finalized to define some Edutainment's methodologies, for an innovative cultural offer, thought to adhere to different cultural contexts.

Keywords: Theatrical mask, Classical theater, body, geometries, Edutainment

1 THE MASK, THE RITE, THE THEATER

The mask is a 'magical tool', charged of values, meanings and functions. The ritual function prevails on the others, and it is very common in many different populations, since it represents a device able to introduce, the person who wears it, into an extraordinary, divine and mystic world. Mask is able to snatch a man far away from his reality, from his contingency, with the aim to project him in a new supernatural dimension. Mask allows, to the one who wears it, to leave his space-time continuum, taking on a new identity.

In the Greek world, during the religious ceremonies, the masks were very popular. From the first references to the animal world in the Demeter's and Artemis's worships, to the ritual masks of Dionysus, god of wine and of the theater; to the giggle masks of the Medusa Gorgon, the monster who petrified with its terrible sight. But the peculiarity of Greek civilization is the use of the theatrical mask that, starting from the second half of the IV century, characterizes the cultural event destined to mark the Occidental history: the born of Tragedy in the name of Dionysus.[1] Therefore, the god of the irrationality and of the euphoria, of the loss of himself and of the excess, of the illusions and of the fluctuation between reality and appearance, between reality and fiction, between male and female, the 'hybrid' god is strictly connected to the theatrical events, to that exhibition feature that puts on act, on the stage, the imaginary as it was real.[2] The visual dimension of the scene is also seated forth by the name of the assigned place: the (da thèaomai, I (can) see), where the architecture builds a prevalent role for the òpsis, the sight, word that in Greek means the seeing's activity, but also the totality of what we are able to see, that is the show. So, the theatrical event had the value of a moral and religious activity, similar to a real rite. Theater was, for Greeks, a mass's exhibition, very popular and cherished by people of every social class or economical condition: it was considered, in fact, as a ritual rich of religious and social relevance, an educational instrument for the community interest. Many people participated to the performances and, probably already during the V century B.C., also women, children and slaves were admitted. Therefore, the theatrical representation is not only a show, but a collective rite of the polis, that is rolled during a holy period in an holy space (usually the theater was build close to a God's altar). The theater, because of its collective feature, assumed the function of a mouthpiece for the ideas, the matters, for the political and cultural life of the community. If it is true, in fact, that Tragedy was conceived to speak about a mythic past, is also true that myth became soon a metaphor for the deep problems that city had.

1.1 *The mask: A medium between real identity and artificial otherness*

The actor's identity is transformed in a fiction's character (the Other) thanks to the suppression of his own identity. In this manner Thespis, the actor, at the beginning decides to cover his visage with white lead and, progressively, to hide it under a mask. On the resetting of the identity, given by the mask, which is obtained when it physically hides the face of the one who wears it, the level of the substitution is dominant, that concerns the different identity that the mask lends to the one who wears it. The mask become a medium, a tool of passage from a real identity to an artificial otherness.

In the relationship that exists between the actor and the character, the use of the mask erases all the actor's somatic features, and it allows him to plunge in that being 'Other from himself", lending the voice and the movement, and alienating from him. On this topic, there

Figure 1. Menander and the new comedy's masks, Princeton.

1. The tradition indicated Thespis as the ancestor of this literary genre. Its origins, according to Aristotle, might be identified in the choral chant for Dionysus. Since 535 B.C., tragedy represents are attested during the springy festivals, officially consecrated during the Dionysian rite as state worship, and legitimated by Pisistratus.
2. Massimo Fusillo, *Il dio ibrido*, 2016.

Figure 2. Prònomos's Krater, National Archaeological Museum, Naples.

is a rich documentation, from the Classical to the Hellenistic's Ages, that depicts the actor during the contemplation of a mask, holding it in front of him before he could wear it, might as penetrating the nature and taking possession of the character's soul. In the Greek theater, mask has two main functions: a ritual function, and a practical function. The connection with Dionysus's worship relates it to the religious sphere, placing it within a rite. Furthermore, the strict convention that limits to three the number of actors constricted to perform many characters, makes necessary its use during the theatrical representations, both tragic and comic.

Mask was defined and completed by costumes. The tragic actor wore stylized and conventional suits; the basic garment was the ketone, a long-sleeves tunic, variously decorated. Also, the costume's color was important, to define the roles and the situations: the red-purple indicated the royal dignity, the black was for the mourning and the misfortune. Also the accessories were important, for example the walking stick for old men, the sword for warriors, the olive's branches for Supplicants, the scepter for kings.

When the theaters, from a first wooden structure, were built using the stone, gradually assume monumental dimensions. Also, the actor's height simultaneously increased, equipping the hairpiece of a high *ònkos*, that is a swelling of the hair on the forehead, and increasing the boots, whose sole, originally short, reached also the 20 cm of high.

Because of the perishability of the materials with which masks was made, any artifact survived. However, there is, a rich iconographic representation on pots and clay artifacts. It is also possible to obtain important information about masks, from the artifacts found in the Greek necropolis of Lipari, in the Aeolian islands, where the digs took back to the high many little decorated theatrical terracotta artifacts, maybe the result of an intense local handcrafted production, tied to the funerary Dionysism, widespread in Sicily and in the whole the Magna-Graecia. Dionysus is also the divinity who gives the beatitude in the afterlife, to those who were initiated to his mysteries. It is possible to set a date for the artifacts around the first decade of the IV century B.C. and the first half of the following century.

Nowadays, these artifacts are exposed in the 'L. Bernabò Brea' Archeological Museum of Lipari, and they represent the widest and completest *corpus* about the topic of the theatrical mask and about the scenic costumes, ever arrived from the Classic antiquity, a real 'material deposit for the theatrical memory'.[3] Thanks to those founds it is possible to retrace almost two centuries of Greek theater: from the Age of the great tragedies of Sophocles and Euripides, until the Menander's New Comedy.

With this theatrical apparatus, discovered by Luigi Bernabò Brea and Madaleine Cavalier, it is possible to admire the historical profundity that increases the material culture of theater, that is the tradition of a civilization that from the V century has been linked to the masks, first of all for a theatrical use, then for a ritual one. Exactly for this second use, it was possible to preserve those artifacts, revealing all the iconographical traits of the Tragedy and of the Comedy's characters of the Classical world. It is such of some theatrical memories, rich of sparks about the acting techniques, and about the side-theatrical activities, obtain

3. In 252-251 B.C., Lipari was destroyed by Romans, and it became an Empire's colony.

Figure 3. (maybe) mask of the Dionysus or Apollo, point clouds. 'L. Bernabò Brea' Regional Archeological Museum, Lipari.

from the little mask's models (for the Tragedy, the Satiric drama, the Comedy, etc.) and from little statues (of comical actors, satyrs, Silenus, auleters, dancers, jugglers, acrobats, deformed characters, etc.), in a gallery that has 44 different types, splitter in five main groups (old men, young men, slaves, old women, young ladies).

1.2 *The mask and the human characters*

The mask's facial features also identified the role and the social state of the interpreted character. For example, the parasite has got a smooth brow, bags under the eyes (sign of a depraved life), evil expression, hooked nose. The 'good young man' has got a regular face and sweet and opened eyes; 'the good girl' has got a pallid skin tone because she lives always inside, decorated with a simple hairdo. The salves have got a big mouth, a lifted eyebrow, snob nose, brown skin tone and hair.

Lipari's masks have been compared with the *Onomastikòn*'s IV book, written by Pollux around the 170 A.C., and dedicated to Emperor Commodus. It is a sort of list of names and synonyms organized on typologies, arranged in ten books. The IV book is entirely dedicated to the masks and to the Greek costume's uses. Although the treatise was written after the Classic period, it represents a rich source of information for the theatrical historians and it gives a complete inventory of the New Comedy 'types' (IV-III centuries A.C.), subdividing them in the forty-four typologies previously specified.

For almost all the types described by Pollux, it was found an accurate comparison with Lipari's masks: 1) the procurer (*Pornòboskos*) looks like the *Lycomédeios*, he has a huge and curly beard, meddlesome behavior, bent lips as in a smirk, joined eyebrows, bald-headed; 2) the perfect young-man (*Pánchrestos neanìskos*) has a reddish skin tone and he is attractive; he has got few wrinkles on the forehead, a crown-hairdo and raised eyebrows; 3) the brunet young-man (*Mèlas neanìskos*) is younger than the others, he has got decreased eyebrows, and he is athletic and polite; 5) the false virgin (*Pseudokóre*) has white face, hair tied around the head and she looks like a bride; 6) the perfect maiden (*Téleion etairikòn*) has a pink skin-tone, lighter than the false virgin and has got ringlets around the ears. This is just a restricted set of samples and types that connect the Aeolian masks with the theatrical classification theorized by Pollux.

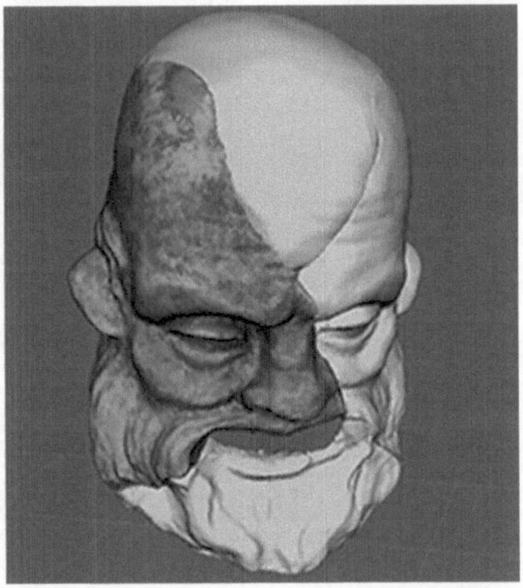

Figure 4. Reintegration of the comic mask of the Pornobòskos, Lipari's collection, Blender.

Figure 5. a) The perfect Hetera's mask, reintegration of the original coloured layer, Blender; b) The main Slave, digital colour reintegration.

2 AMONG ANTIQUE ARTISAN METHODOLOGIES AND MODERN DIGITAL PROTOCOLS

2.1 *Differences between the theatrical mask and the ritual mask*

In this context we developed a three-years research about the material culture of the theater, examined though the collections of artifacts, stored in the Lipari's Museum. This research is based on meticulous bibliographical and iconographical references and on an accurate digital survey of the masks.[4] As first, we marked the difference between the theatrical mask and the ritual mask.

Although both of them reproduce the features of the same characters, the first one is placed in a specific ludic and pedagogical area, at the service of the wide textual repertoire of Tragedies and Comedies of the Classic theater. The ritual mask fulfills to the Dionysus's rites connected to death and for this reason was found in many samples and size, especially in

4. This Research was developed during a three-year Ph.D. Thesis in '*Architettura e Territorio*', with the title of 'The Material Culture of the Theatre: study and analysis of the Liparote theatrical masks of the Magno-Greek *coroplastica* for the compilation of a (digital) protocol for their integration and anastylosis, Ph.D. Andrea Marraffa, relator Full Professor Francesca Fatta.

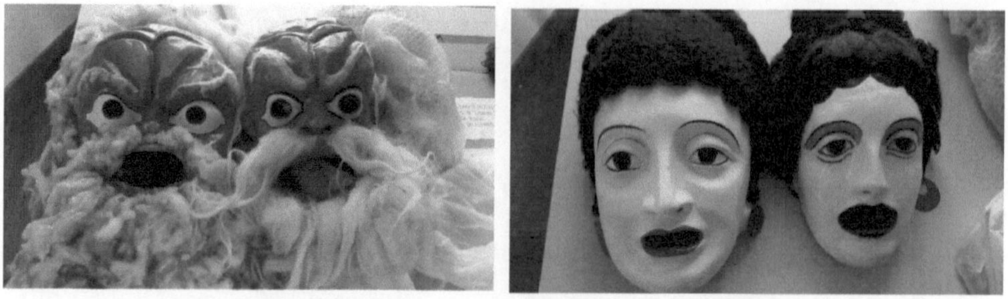

Figure 6a–b. The Prototypes of four comic masks for the contemporary theatre (two old men and two young ladies), R. Mortellaro.

the Lipari's necropolis. The theatrical mask had specific ergonomic features, like for example the lightness, because it had to perfectly adapt to the face's shape of the actor. He had the necessity to maintain a visual and acoustical filter through the appropriate positions of the eye's and mouth's openings. The masks had to be able to show, also from a certain distance, the main features of the characters, and for this reason, the colors were well saturated, with luxuriant wig. Because of the perishability of the materials, as already said, we do not have any original mask, but only images treated from pots, little statues or mosaic decorations. The ritual mask was made with clay and its dimension was tinier than the normal size. It has not any holes for the eyes and, often, neither for the mouth, because they did not have to be worn. The colors, that we can admire, on the base on the many artifacts, confirm the sharpness the ancient theatrical masks once had.

2.2 *From the photogrammetric survey of the artifacts to the digital protocols for the reconstruction and the anastylosis of the masks' fragments*

Nowadays the use of modern digital instruments seems to be an indispensable practice for the knowledge, documentation and dissemination of the Cultural Heritage. Therefore, once we defined the preliminary and necessary theoretical state of the art about the theatrical masks of the Magno-Greek *coroplastica*, we thought about the possibility to draw up some methodologies, first of all, for the direct survey of artifacts, then, for their subsequent phases of digital reconstruction and reintegration. The Lipari's theatrical masks may be schematized, from a morphological point of view, in three distinct degrees: entire masks, fragments of entire and simple fragments or 'mute'. The workflow of reconstruction and digital anastylosis, therefore, respects the decomposition of these three degrees of *status*. During all the pipeline of work, we also tried to underline that constant relationship, almost instantaneous, between the identification of the problem and its immediate resolution, which is bedrock of the 'doing's' methodology. A continuous 'flow of planning circularity', that is able to combine manual and digital practices. It is, in other words, to suggest what R. Sennett defined as 'the eternal rhythm that is created when the artisanal experimentation remodels the matter'. The continuous advancing and returning during the various phases of work, has always been the sacred fire that nourishes the relationship among the project, the 'transformative material activity' and the final product. Undoubtedly the alternation between thinking and doing was very important for the material activity of the *coroplasticos*, the master-craftsmen who produced our theatrical masks. Therefore, during the digital workflow of reconstruction of the artifact's fragments, we tried to reiterate the manual nature, own of those ancient processes: a modern methodology but, conceptually, very ancient. The *leitmotiv* of the protocol was, therefore, to 'think as artisans', making a strategic use of the innovative digital techniques. The protocol has a reverse pyramidal structure which from the general, goes until the detail, gradually tightening the focal to an ever more specific part.

If the process of integration and reconstruction of the artifacts has been applied to the most representative terracotta simples, and belonging to all kinds of drama – Tragedy,

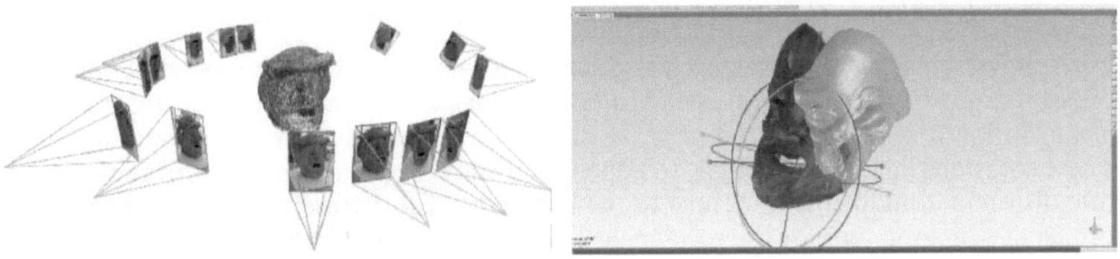

Figure 7. a) 'tout autour' methodology for the digital photogrammetric survey, mask of the 'Curly young man', Lipari; b) Pornobòskos, digital 3d modelling and sculpting, Geomagic Studio e Maya.

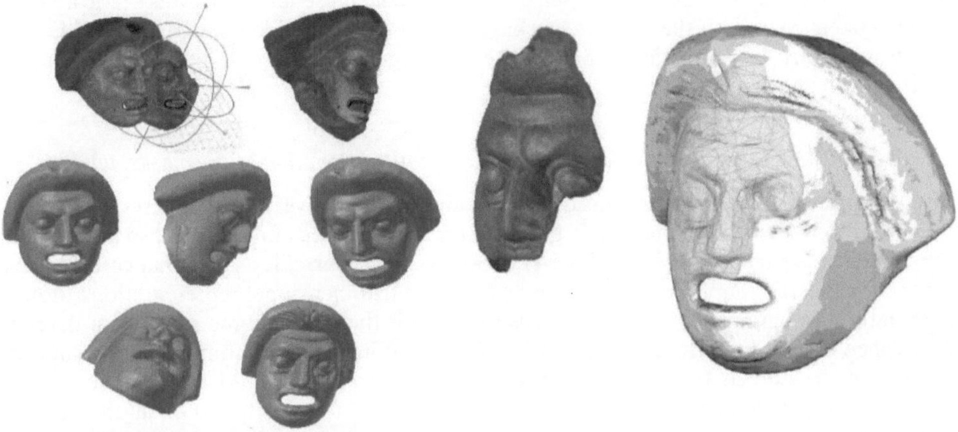

Figure 8. a) Generic prototype's modelling phases; b) anastylosis's methodology and adherence's testing of the 'mute' fragments on the generic prototype.

Ancient and New Comedy, Satyr Drama – on the other side, the pipeline of anastylosis it mainly focused on the New Comedy's masks. These masks, in fact, have been discovered in Lipari with a greater frequency than those of all other theatrical genres. In addition, the subdivision in macro-groups of comic masks (old men, young men, women, slaves) and in sub-groups (for example: the *Pappos pròtos*, the vainglorious young man, the parasite, etc.), allowed to orient the protocol to those particular categories that present a higher number of specimens. So, recapping, the first part of the Research deals with all the adopted digital phases to reconstruct a mask, starting from its state of 'fragment of an entire'. In particular, the workflow is articulated through the following phases: – photogrammetric survey; – data processing (extraction of point clouds); – meshing of models; – study of the main geometries; – philological and bibliographical detailed study for the two-dimensional reconstructive hypothesis; – 3D modeling (sculpting and texturing phases); – texturing, neuter zone criteria and evaluation of the color dominances. Instead, for the anastylosis protocol, after a first general reconnaissance of all mask groups, starting from the 2D redesign of all the 343 masks New Comedy, we decided to reduce the focus towards the macro-group of young-men, and subsequently on the subgroup n. 12, the so-called 'Young men with wavy hair'. On this subgroup – and in particular on the specimens of the second measure – we were able to define an anastylosis pipeline, with the main aim to build a linguistic code useful to relocate all mute fragments within a larger compositional scheme, able to including all. In particular, the anastylosis is based on the definition of a general prototype, which is a neutral and generic model, extracted by the weighted average of all measurements drawn on sub-group specimens. Finally on this prototype, we verified the adherence of the mute fragments and also many other parallel considerations useful for the philological and semantic analysis of the finds.

2.3 *The edutainment: The didactic entertainment between mask and body*

The following part of this Research is oriented to the definition of some methodologies of Edutainment, applied to the Magno-Greek mask's universe. The word Edutainment means a liquid kind of cultural offer, at halfway between education and entertainment. These solutions have been thought to adhere to different cultural contexts: from the scientific dissemination for the museums to the recreational activities for the theater, from the choice of semantic tool for the pedagogical and didactic valorization to the studies about the design and the ergonomics of our theatrical devices, finalized to their modern reuse in the theater.

In particular, we tried to underline: the study about innovative solution for the segmentation and semantic annotation of some scenic devices; 2) the 3D prototyping of some mask's model for the design of a real 'Multi-sensorial and tactile museum of the Mask'; the analysis of AR systems based on the Motion tracking and on the Face-video mapping, with the aim to develop a series of activities with a precise pedagogical orientation, for the museums and for the theaters.

2.4 *Methodologies for the annotation and the semantic description of the masks' universe*

With the aim to guarantee a semantic in-depth analysis of the very wide Lipari's mask's universe, we decided to use an innovative platform for the semantic annotation, also know Aïoli. This software, tested at the CNRS's MAP laboratory of Marseilles, is a cloud collaborative service for the 3D annotation, that allows to process tridimensional representations though automatic photogrammetric tools, and also to enrich them with some semantical descriptors, called denominations or labeling, with the aim to improve the comprehension and the analysis of the studied artifacts. The software was created to make more innovative, flexible and shared the documentation of the Cultural Heritage. Aïoli was tested on all the samples of masks (entires, fragments of entire and 'mute' fragments) of the New Comedy. We defined a hierarchical schema of calques, and we draw up so many layers how they were the semantic descriptors to link. For each sample of mask we extract many information about the morphological and geometrical features of the artifacts, functional for the qualitative and quantitative comparison between a sample and the other. Among all these information, we may mention: complex vertical profiles, bending radius of the main elements (mouth, eyebrows, cheeks, nose, eye), quantitative analysis of the primary elements comported to the entirety of the artifacts, main distances, etc. Starting from a precise shape grammar, constituted by all the primary characters, remarkable on each model, we were able to define a sort of hierarchical clustering for the macro-typologies of masks and of their main morphological elements. This typology of structure – also called 'tree-system' – is a phyllo-genetic structure, and it leads all the semantic description of each mask's samples. The mythology of the semantic description helped us to have a topological control in each step during the entire management of the models. It showed us how it is possible to define a continuous workflow of information, codified in one only coordinated and coherent source. This methodology, with a huge technological pregnancy, but at the same time equipped with a strong artisanal character, allows to obtain real 'talking' models, thanks to their multi-resolutions. The labeling and the

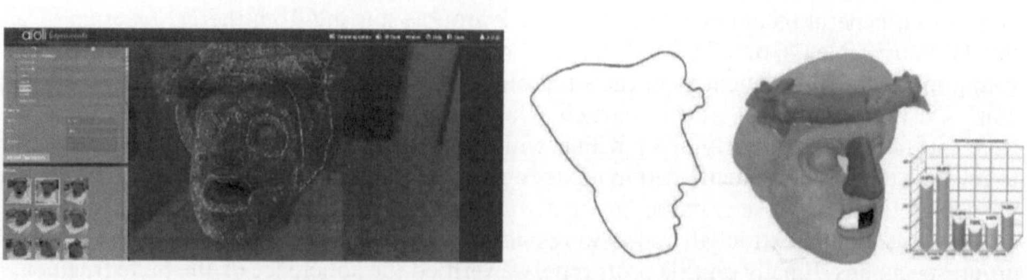

Figure 9a–b. Main information extracted from Aoli, vertical complex profiles and quantitatively analysis of the primary elements reported to the entire artifact.

semantic description seems as a necessary process for the Cultural Heritage's documentation, and they represent an indeclinable approach for the completest and widest communication and dissemination of the Cultural Heritage.

3 FROM 3D PROTOTYPING TO THE TACTILE MUSEUM

The three-dimensional rapid prototyping of the Lipari's theatrical masks, discloses the last part of the methodological protocol presented in this research. With the aim to define a tactile and multi-sensorial path, available for everyone (users with or without visual disability), we printed some of the most significant artifact of Lipari's collection. Among them, we chose the sample that better than others condensed all the step developed during the protocol of reconstruction and anastylosis. Therefore, we printed some entire sample, some fragments of entire, and some simple or 'mute' fragments. It is important to underline that the design of a tactile and multi-sensorial museum does not aspire to replace the traditional cultural offer. Quite the opposite, the main objective is to juxtapose to the current collection, increasing it through some new more interactive and captivating systems of fruition and dissemination. The rapid prototyping step implies, upstream, the phases of survey and of parametrical modeling, presented in the precedent paragraph. The digital photogrammetry allowed to have high density and high-resolution models, and it also allows to respect the original artifact's morphology, giving a precise reading of each part. About the model's printing, we used 3D *ProJet 660Pro* color printer, with plaster powder and synthetic resin, that works for addition of layer. This printer is able to reproduce high resolution models, with all those color and texturing information acquired during the scanning process.

From a side, the three-dimensional printing allows to have models with a high morphological and formal accuracy, however on the other side, this does not work as well for the chromatic yield. With the aim to guarantee a more accurate hicastyc mimesis of the original artifact, we made many color testing, printing at the same time many copies of the same model, but with a reduced size. All the textures, extracted from the photogrammetric survey, undergo some post-processing modifications, through the application of filters, from time to time, warmer or colder than the original texture. Then, each texture was applied on many copies of the same 3D models. So, each artifact was compared with the original one, and we chose the copy that, more than the others, looked like the original one. We printed this copy at the real 1:1 size. We cleaned it from the surplus powder and we treated it with a particular acid resin, which made brighter and sharper the colors. It is good to remark two important aspects: – some models, in particular the fragments of entire of the Dionysus and of the Hethera, were printed in one single-piece, with the relative part rebuilt whose we applied an under-tone of color, with the aim to guarantee the recognition of the integration. For the other fragments of entire, for example in the case of the Pornòboskos, we decided to print two separated pieces: one for the original fragment and the second one for the related integration. We also thought about a magnetic system to make possible the coupling of the reconstructed part and the original fragment. This aspect guarantees the independence of the fragments in relation to the entire, and at the same time it allows immediately to return to the unity. In this way, we respected two important theoretical objectives for the restoration: the recognition of the intervention and the reversibility of the reconstruction process; – the color printing is, for obvious reasons, useless when the models are touched by a blind or a sightless audience. For this reason, with the aim to simplify the fruition and the recognition of the original parts in relation to those rebuilt, we thought to operate on the texture's grain: smooth for the rebuilt part, rough the original one. Moreover, the two parts were separated by a tiny dividing line. For what that concerns the overall set-up of the tactile path, this one is almost similar to the current cultural offer, but it increases it adding a series of handrails with particular metallic incisions. These mark the time and indicates the presence of some special shelfs where the printed artifacts are collocated. Finally, some three-dimensional panels, tactile maps and Braille signs, briefly illustrate the mask.

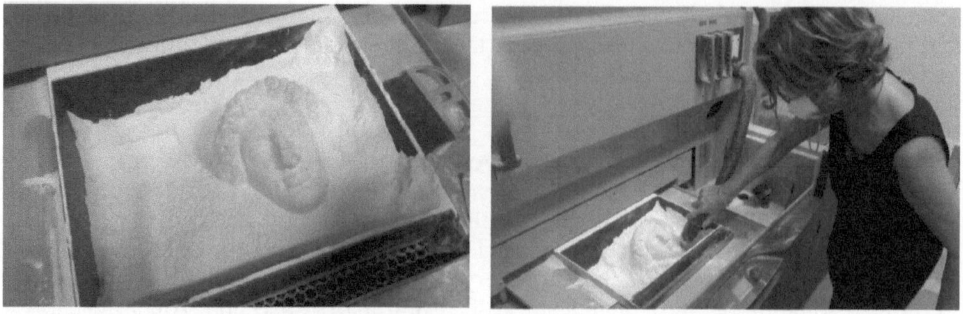

Figure 10a–b. 3D colour printing of a comic Lipari' mask, Dionysus or Apollo.

4 THE AUGMENTED REALITY AND THE NEW DIGITAL MEDIA. TWO SOLUTIONS FOR THE MUSEUM AND FOR THE THEATRICAL CONTEMPORARY SCENE: THE MOTION TRACKING AND THE FACE VIDEO MAPPING

How we already said in the previous paragraphs, the cultural spaces became, always more, a complex business, operating within a new cultural economy that is in constantly evolution. Nowadays, the Cultural Heritage's business is variegated but also very saturated, because it is still linked to a too slow and passive communication, unable to encourage the final customer's participation. There are still few museums that choose to wager on the digital way with the aim to integrate the traditional offer, setting up all the modern and interactive technologies to create new story-telling for a more complete dissemination of the Cultural Heritage.

For all these reasons, near the creation of a multi-sensorial and tactile path of the mask, described in the previous paragraph, we also thought some other Augmented Reality's solutions, condensed in the Motion Tracking and in the Face-Video Mapping, with the aim to guarantee a faster and a more captivating fruition of the analyzed artifacts, customizing and regulating the visit on the base of the necessities and interests of each visitor. We developed these AR solutions to respond to multiple contexts, characterized by different declinations: form the classical dissemination for the museums, to the didactic for the academicals-scientific area, from the serious gaming to the fruition for the show business, the entertainment and the theatrical and architectural scene. The AR solution for the museums have been also developed to guarantee an easiest comprehension of the reintegration and anastylosis's digital protocols. In particular, nearby the artifact's showcases, we thought to arrange two screens connected to four depth-detectors. So, the user can interact, with a movement sensor system, able to scan and trace the human face. One it obtains the tracking of the visage, the user has the possibility to select, through a specific drop menu and in an interactive way, the typology of mask that he prefers, from a pre-charged model's pre-set. The masks are divided into some categories, for each theatrical genre (Tragedy, Ancient and New Comedy, Satiric Drama), and on the base of the group and of the sub-group whose they belong. On the high right side of the screen, a short text describes the masks and an interactive map indicates its position in the museum. Once selected, the mask is scaled and oriented, therefore, accurately calibrated on the user's head.

He has, in addition, the possibility to interact and to interpose himself inside a hybrid augmented reality, choosing among different scenarios, for example some existent theatrical scenes or digital reconstructed theater.

This aspect allows him to understand what could be the real perception of this scenic device, in the ancient theatrical contexts, by means for example, a simulation of the vocal amplification, though a rapid comparison of the device's morphology and size, one wore by the user, and the relationship between the device and the virtual theatrical scene. The calibration is based on precise ergonomic analysis of the entire masks at the real size, done to comprehend how a human head may adapt inside it. All those operation here presented are

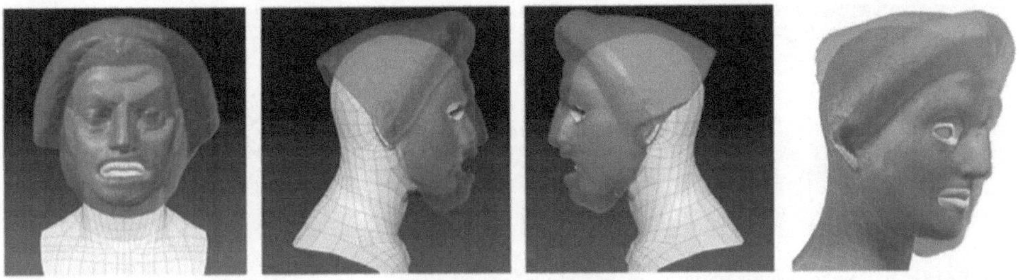

Figure 11a–b. Ergonomic studies of the generic prototype's mask, real human size.

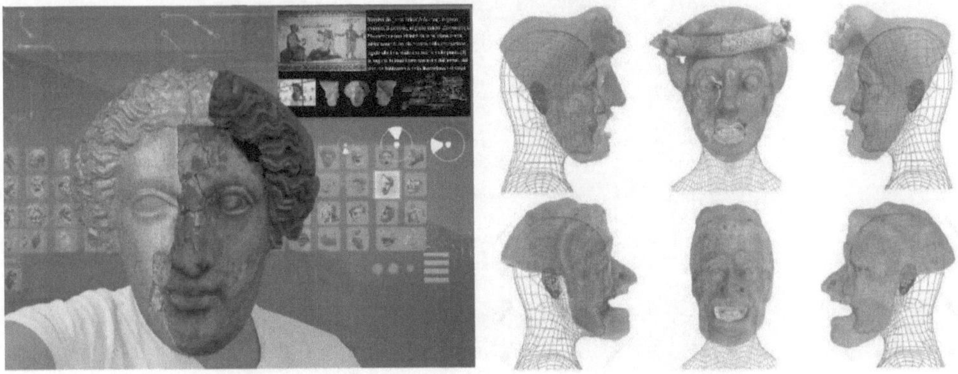

Figure 12. a) Selection of the Dionysus's mask from a pre-set menu of available masks, motion tracking, Edutainment for the museums; b) Ergonomic studies of the Young man's and of Hérakles' masks, real sizes.

available for the detected entire masks. Instead, for the fragments, (fragments of entire and simple or 'mute' fragments), the interface offers the possibility to select a reconstructed mask, and also to select the relative, neutral and generic prototype on which it is possible to arrange the corresponding 'mute' fragment, on the base of the fundamentals before described into the protocol's pipeline (anastylosis).

The Face-Hacking is another technique for the strengthening of the reality's perception, this time thought for the theater and for the modern scenic representations. This methodology is based on the projection (mapping) of images and videos of physical real support. The mapping can be static or dynamic, and the surface of which the images is projected, can be a simple three-dimensional object, or also a face. The methodology was thought with the aim to reintroduce again the mask in the contemporary theatrical contexts. Today, the mask appears as a secondary scenic object, almost completely ruled out, or restrict to some brief moment during the drama representations. This condition is probably due to the fact that this device, carrying out the main function to change the actor's face, transforming it in an 'Other from himself', excessively limits the facial miming of who is wearing it. Therefore, it seemed necessary find an arrangement able to readmit the mask in the contemporary dramaturgy, and on the other side to guarantee at the same time the recognition of the actor, the miming and the facial expressivity during the acting. The Face-Hacking represents the ideal tool to achieve these objectives. In this way, the mask is able to become a sort of second skin for the actor, perfectly overlapping to every single facial muscle. This technique, as we already mentioned, is based on the real time projection of the images, directly projected on the actor's visage. The good success of the Face-Hacking depends from the correct correspondence between the pre-set images and the face, but also from the abilities of the actor to dynamic interact with them. Also, in this case the face has been traced thanks to the application of specific white

Figure 13a–b. Face-tracking/mapping and morphing of the Hethera mask and of the Dionysus mask in a real ancient theatrical context, Adobe After Effects and Blender.

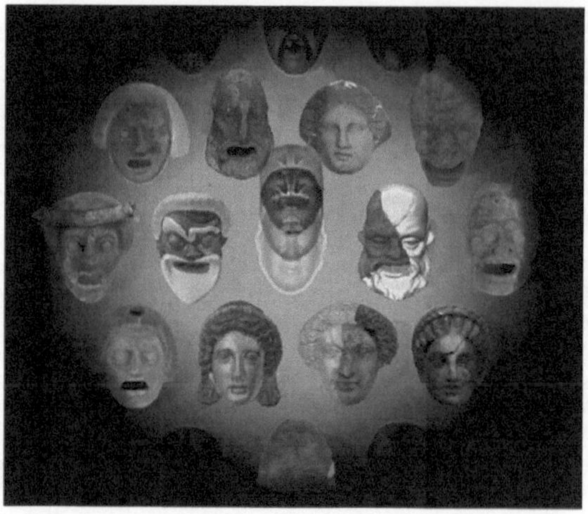

Figure 14. The mask's wall, video mapping application, for the dissemination of the research.

and low-contrast markers, distributed on all the zone of the visage: front, eyes, nose, lips, chin, etc. More are the markers, and higher will be the adherence and the expressive fidelity during the projections. Once we traced the face, we try to associate to the entire composition, the digital masks. We tested these workflows mainly on the masks of the 'Perfect Hetaera', of the 'Fake virgin' and of the Dionysus. Every images and digital models has been calibrated on the faces, by means the overlapping of a polygonal plot (marionette plot. Then every facial movement has been recorder on a specific time-line, so it has been projected on the mask (filet). Finally, through different masking operation, we created the eyes and mouth contours, preservation the original part of the actor, with the aim to guarantee the real expressivity of acting during the projection.

5 CONCLUSIONS

The main objective of the Research was to study analytically a large part of the collection of Lipari's masks, with the aim to draw up a protocol of integration, anastylosis, and of dissemination about the most representative specimens. The study has also allowed to arrive to further specific conclusions: 1) the first, in a theoretical sense, asserts that it is possible to talk about the existence of a Material Culture of the Theater. It is an essential part of the theatrical world. Its presence is crucial to triggering all those mechanisms of sewing between the tangible and the intangible culture in the scenic arts world; 2) it has been defined a complete and universally workable pipeline, applicable to all the fragments of entire and to all the 'mute' fragments. It considered all the main cases and the problems that the fragmentary state involves. As we already explained, this protocol is potentially applicable also to other areas, different from those of the archeological area. In fact, the methodology, which follows the logical line of 'digital scanning of the finds/modeling by means of integration/construction of the anastylosis pathway' may be conducted, for example, even on fragments of ruined architectures, for which it is possible to reestablish a historical lost memory. In addition, the segmentation and semantic description steps are particularly indicated when it comes to deal with complex assets of Cultural Heritage, for which it is necessary starting a schematization in hierarchical structure or into clusters, progressively more sectorial and specific; 4) last but not the least, we developed some innovative way useful for the dissemination, the fruition and the valorization of the scenic devise, for their reusing and overhaul in the modern museums and in the contemporary theaters, through virtuous, innovative and interactive approached of Edutainment.

REFERENCES

Barra Bagnasco, M. 1986. *Protomi in terracotta da Locri Epizefiri*. Torino: Il Quadrante, pp. 20–23.

Bauman, Z. 2012. *Liquid Modernity*. It. Trans. *Modernità liquida*, 2a ed. con nuova introduzione, Roma–Bari, Laterza.

Bernabò Brea, L., Cavalier, M. 2001. *Maschere e personaggi del teatro greco nelle terrecotte liparesi*, Roma: L'ERMA di Bretschneider, pp. 18–19.

Bernabò Brea, L. 1981. *Menandro e il teatro greco nelle terrecotte liparesi*, Genova: Sagep Editrice, p. 7, pp. 11–14, pp. 21–27, p. 36, pp. 120, p. 146, pp. 154–157, pp. 159–160, p. 169.

Bibier, M., 1961. *The history of the Greek and Roman Theater*. New Jersey: Princeton University Press, pp. 28–36, p. 53, p. 81.

De Luca, L. 2011. *La digitalizzazione del patrimonio costruito. La foto-modellazione architettonica*. Palermo: Dario Flaccovio, p. 5, p. 24, p. 34.

Manuel, A. 2016. *Annotation sémantique 2D/3D d'images spatialisées pour la documentation et l'analyse d'objets patrimoniaux*. Aix-en-Provence: ParisTech, p. 26, p. 97.

Vernant, J.P. 1987. *La morte negli occhi. Figure dell'Altro nell'antica Grecia*. Bologna: Il Mulino, p. 45–57, p. 62.

Wiles, D. 2007. *Mask and Performance in Greek Tragedy: from Ancient Festival to Modern Experimentation*. London: Cambridge University Press, pp. 33–35, pp. 237–260.

Drawing (...) City (...) Body, Dwelling on Earth – Janeiro (Ed.)
© 2019 Taylor & Francis Group, London, ISBN 978-0-367-02602-8

Cinema's places and territories: A systematic catalogue of Italian movie theaters

F. Moschini
Segretario Generale dell'Accademia Nazionale di San Luca, Ordinario di Storia dell'Architettura, A.A.M. Architettura Arte Moderna, Italy

F. Maggiore
Presidente della Fondazione Gianfranco Dioguardi, Direttore Creativo di Kiasmo, Italy

V. D'Alba
Architetto, illustrator e designer di Kiasmo, Italy

ABSTRACT: Cinema halls represent a vital asset in the cultural offering of any region. Being aware of this value is indispensable in planning their development and enhancement. This premise led to the need for a survey of movie theatres in Apulia, Italy so as to constitute a data bank of information concerning their present condition, in order to be able to programme their future. The research conducted for Territori del Cinema will lead to a proposal for development, distances between establishments, and the cultural offer of cinema, theatre and entertainment venues. Historical and literary evaluations, sociological and statistical considerations, together with architectural and infrastructural proposals, will help to wisely define the renovation and redevelopment of Apulia's cinema network. Hopefully this survey will stimulate thought on a wider cultural operation, reaching out from Apulia to include neighbouring regions and eventually involve the whole country.

1 INTRODUCTION

Italy has been affected by a progressive impoverishment of the architectural and film heritage. Since its inception, through the most significant moments of the thirties and the fifties, until the spread of new media, cinema as a place has always been playing a leading role within the urban and cultural city context. In the most attractive city areas and in the suburbs, cinemas have firstly portrayed places of life, where the events in memory were not often those seen on the screen.

Screens are not only the boundary lines of the theatre auditorium, but also the surface on which the urban reality can turn itself into image, replacing the completeness of its geometrical

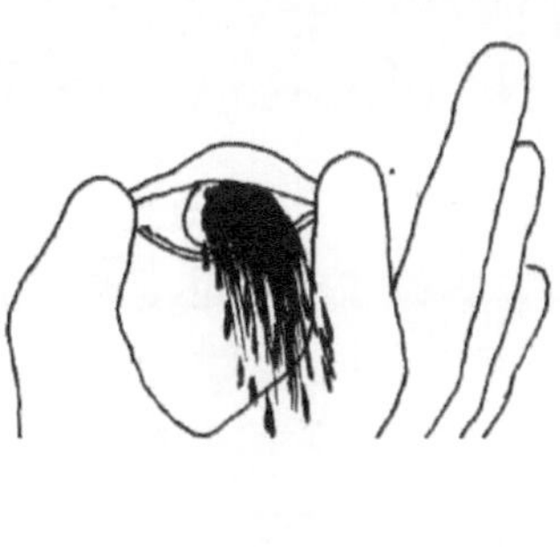

Figure 1.

representation with the partial fragmentation of the urban vision caught in the individual's aesthetic experience living in the city according to its organization and rhythms. Similarly, theatre is not a Cartesian empty space, but a field strength, the condensation of the dynamic interpersonal interactions that occur in the city, in the crowd, revealing, therefore, a natural continuation of the street. Despite the noble function that cinema holds, since its origins, nowadays there is a growing decay of the assets of the existing structures which are subject to user changes thanks to high degree of transformation and adaptability.

This particular feature has led to the consolidation of the cinema metamorphosis; theatre is becoming able to adapt itself to the diverse types of environment and functions far from the entertainment world and closer to supermarkets, garages or discos, even to car parks, banks and gyms. Those solutions are primarily derived from the architectural features of the buildings that have a grand entrance, between the urban pattern and other independent buildings. The extreme metamorphosis can determine, most of times, the disappearance of suitable places for cinemas: places for cultural development, places for a form of escapism, night places, public places, abstract and empty spaces, as important as squares and, therefore, the few remaining places that give historical continuity and significance to the city.

Therefore, people's awareness of the cinema heritage is becoming essential, as well as a comprehensive action of mapping and cataloguing Italian cinemas and theatres, beginning with the research project "Cinema territories: rooms, places, landscapes. A system for Apulia", promoted by the Apulia Region, A.A.M. Architecture Modern Art and the Polytechnic of Bari. The mapping of the Apulia cinemas has been carried out throughout the region by means of different sources and research tools.

First of all, the "Running Cinemas" and the "Cinema Register" were provided respectively by IGES (Italian General Entertainment Association) and NCEA (National Cinema Exhibitors Association). The census was widely performed through field investigations, such as police stations, municipalities, operators and owners of every single theatre. The survey was divided by province and carried out by collecting information and technical data for each facility: location, type, year of construction, features, designers, number of rooms and seats. All theatres were included in a geo-database through the acquisition of geographical position in UTM coordinates (Universal Transverse Mercator) – WGS 84 zone 33. As first result, it

Figure 2.

Figure 3.

produced a map of the distribution of the facilities in Apulia, which provided information about all theatres. 269 structures were identified in single-screen cinemas, multi-screen cinemas, arenas, multiplexes and drive-in cinemas of which 144 running and 125 inactive. The distribution of running and inactive theatres on the region is not homogeneous. It is conditioned by several factors: the topography of the area, the distribution of the municipalities in their respective provinces, their accessibility, population density.

By drawing up the census there came a faithful and detailed picture about the architectural and structural characteristics, history and anecdotes about each detected theatre. All files identified with a code show key information (name, location, address, inaugural year, property, architect, number of rooms, number of seats, type of system and programmes) on one side, historical, analytical and graphic descriptions on the other one. The files have been drawn by consulting documentary material at Municipal Technical Offices, Archives, land registers, public libraries and private collections. Research has led to the collection of large quantities of documents: drawings, tables, project reports, testimonies, postcards and historic photographs. Each file, in addition to a photographic survey carried out by the authors, which gives account of the current state of each analyzed cinema, makes use of authorial reading of thirty photographers, involved in order to explain all the cinemas in a personal and artistic style. Most of the material collected and used for the cataloguing of the cinemas is almost totally unknown. However, it is significant that both the dispersion of many original documents and the fragmentation of news may have caused gaps and inaccuracies.

The centrality, even symbolic, that single-screen cinemas once embodied has been tempered by the multiplicity of existing movie theatres, but on the same time strengthened and improved in ways that have not simply replaced old ones but were born in completely different ways and places.

These large buildings called multi-screens and multiplexes are distinguished by the ability to offer the viewer a wide choice of films as well as extending the definition of cinema, so far referred to as a single place, an identity and historical context within the city. They have gained another urban and architectural meaning especially in settlement patterns and the increasingly numerous and varied contents, so now representative of a well-established part of the city.

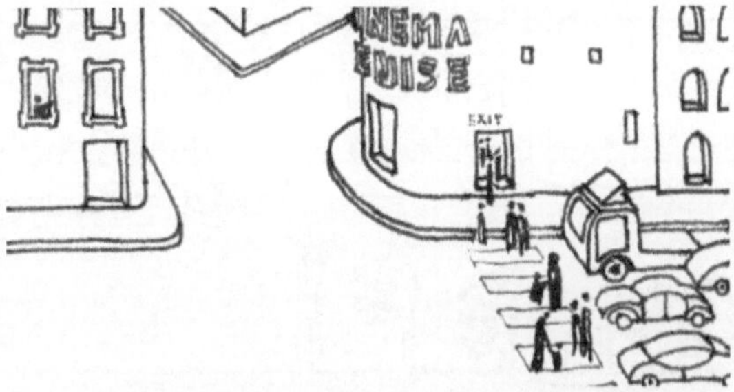

Figure 4.

Figure 5.

A "single-screen" is a building or a place for public shows, with one screen and equipment needed for films. The term "single-screen" spreads following the birth of the multiplex; until then, the word "cinema" was popular. The single-screen cinemas is the first type of entertainment developed simultaneously with the birth of cinema. Its formal features have been transforming over time: initially cinematic representations were through vendors and temporary exhibitions housed by public and private settings; due to the decline of the errant cinemas, after the first decade of the twentieth century there spread the fixed structures, by installing them within existing theatres or buildings. Since the twenties there has been a growing structural, technological and formal evolution. Single-screens have suffered various aspects: isolated building; local parish; a place enclosed into residential buildings. They are often placed in strategic locations in city centres.

From the historical point of view it is essential to observe how the structure of single-screens since the time of their maturity until the introduction of multiplexes has been kept unchanged, beyond the various stylistic changes. On the contrary, the multiplex system, creating greater complexity, has been constantly evolving both in its architectural form and in the perception of society which has become more and more willing to recognize the changes. In fact, if on the one hand there are numerous protests to raise awareness about the protection of single-screen cinemas as a cultural and architectural heritage, on the other one the multiplex is now accepted as an essential value of urban innovation. However, the historical cinema heritage has been decaying. The difficulty of finding new use, despite the existence of examples of undisputed architectural value, is one of the main causes for such a situation.

Even though a number of cinemas have stopped running, entertainment and services have been improved by means of technological renovation of the buildings and more comfort. Cinemas have been renewed as places of socialisation, aggregation, and culture on the territory that, as such, deserve to be appreciated, protected and revitalized. Single-screen cinemas both develop cultural opportunities of individual municipalities and allow parallel sectors to survive on the market. Together with the support of cinemas by the "participants" of the market and the system there might be the use of digital technology and satellite broadcasting;

Figure 6.

Figure 7.

so that cinema becomes a promoter of itself and of its offer, increasing variety and flexibility in programmes and choices.

The crisis in the cinematographic industry, together with the changes occurring in consumer mechanisms, mobility, customs, is to be seen in the transformation of cinemas, which coincides with the birth of the first multi-screens and multiplexes later. These new centres of consumption, mostly common in suburban areas, help more than anything else to the redefinition of the contemporary urban landscape since they are now the new centres that over time will improve their functions, standing for trade centres, meeting places and leisure centres. The origins of multiplex are better understood by another step. Indeed, in the seventies the first attempt to address the commercial crisis was accomplished by splitting a traditional theatre in a large number of small theatres. That phenomenon gave way to "multiboxes" that soon entered into crisis, because of lowering comfort standards after architectural work. First, the problem of soundproofing between rooms and secondly too close seats contributed to multiboxes desertion and multiplex design to show more than one film. In the eighties, in Europe, multiplexes included at least eight screens. Today these structures are identified and characterized by the presence of car parks, optional services, shopping centres, with high quality standards.

A "multi-screen" is meant to be a structure containing two or more theatres used as multiple show schedules with a single ticket office and other services, such as bookshops, shops,coffee bars, restaurants, arcades, exhibition halls. This type of cinema is to be found in city centres, after the splitting up of the old single-screen cinemas, and in the suburbs where it usually comes as a new design. Multiplexes mainly developed in the United States in the sixties, while in Europe they started spreading in the eighties. They were increasingly built in suburban areas and in shopping centres. The multiplex formed by 2 or 4 screens was a result of the division in more theatres; a very popular solution, applied to large single-screen cinemas, with stalls area, gallery and stage. That type of intervention did not involve any alterations, and the audience was in the main room, while the gallery and the stage were in other rooms.

In recent years, people involved in the analysis of urban and local alterations have been trying to understand the meaning of multiplexes in relation to existing public spaces in an area. Two types of cinematographic architectural forms have been identified: one according to the various components enclosed in a single structure, the other according to an autonomous

Figure 8.

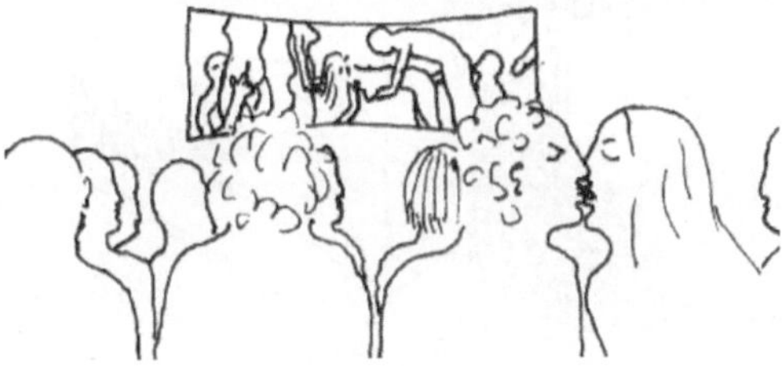

Figure 9.

cinema structure, as well as other services, all in close proximity to each other so as to combine and connect the audience's flow.

Among the most common definitions of "Multiplex" there was one proposed by Media Salles, according to which the multiplex was a structure with 8 screens at least. This proposal was confirmed by the results of the study carried out by London Economics, published in the "White Book of the European Exhibition Industry" in 1994 which asserted that "the multiplex effect is not fully achieved unless there are at least 8 screens." The definition itself has been also approved by the CNC (Centre National de la Cinématographie).

The first multiplex was opened in July 1963, at the Ward Parkway Shopping Centre in Kansas City, Missouri. Multiplexes have similar architectural features to shopping centres, mostly without formal qualities. These complexes are located, indeed, almost exclusively in the suburbs where large car parks are needed. The layout of the interior rooms includes a large foyer, together with a series of spaces for people's rest, refreshment and entertainment. The arrangement of the rooms usually has a symmetrical system which improves the use of space, providing a central corridor for all rooms. The projectors are set within a single projection box. The word multiplex implies various qualities: the most important and influential is the urban one. Cinema producers aim at offering greater quality, therefore they keep investing in multiplexes. Traditional single-screen cinemas then appear to be in danger especially in big cities. The construction of a multiplex cinema inevitably implies proportionally lower operating costs than a single-screen, less risk compared to product demand which is rarely predictable, better marketing strategies and the right to neutralize competition showing films in more theatres ("saturation selling").

In the nineties mega-plexes seemed to put an end to the rapidly increasing growth of old cinemas: structures had sixteen to thirty screens and based their commercial politics on quality and versatility. The chance to select a film out of different ones in new multiplexes made cinema closer to television, where selection was varied still remaining in the same place. Old single-screen cinemas tried to survive by turning themselves into multiplexes in order to update according to the audience's needs and at the same time to become adapted to new technologies. The ability to manage the planning in different theatres allowed the distribution of different films: American films and art-house cinema. Hence the introduction of

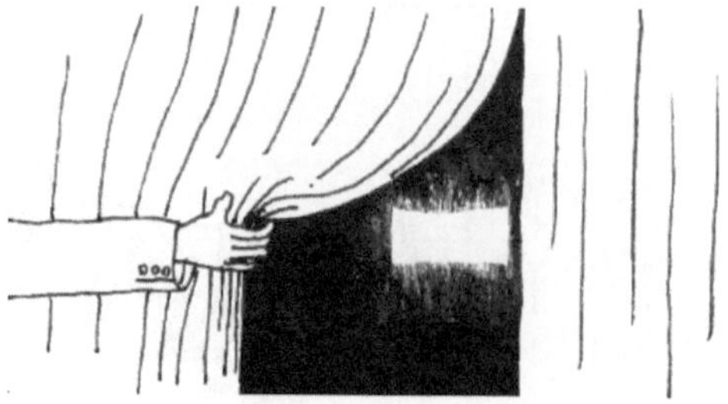

Figure 10.

Figure 11.

foyer which, in spite of the commercial approach, represented the meeting place of the mass of users. In this context "multiplex" could be properly defined as a film structure with clear peculiarities, such as good accessibility, i.e. proximity to large arteries or road junctions; variety of film programming, i.e. a wide selection of products, because of theatres and schedules; high-performance technology, because of visual shows and high-quality sound; the presence of attractive complementary services aiming at entertainment, commerce, and leisure. In the past people used to go to the cinema to see a film and most of times they used to go back home after the show. Nowadays multiplex is one of the many entertainment venues, where time is lengthened in shopping centres. In recent years the economic role of multiplexes has established itself in an even greater way.

These new centres, based on commercial choices, should also respond to qualitative and architectural requirements. Indeed, the selected area, namely Apulia, in spite of the considerable number of cinemas, shows an idiosyncrasy; cinemas are seldom an added value to the context, since they are considered as functional containers. Moreover, those few public works worthy of architectural potentialities are generally the results of anonymous building commissions. However, the anomalous situation contains models, references, images related to the architectural production outside the region, that is in the rest of Italy.

When single-screen cinemas closed, there was an increase in the number of screens caused by the opening of multiplexes; it followed that the management and the distribution of films were in the hands of a few operators. The growth of multiplex structures at the expense of the small cinemas should have been however more regulated by legislative and policy instruments. The continuing economic crisis and the increasingly poor film quality prove these data to be right, but it is clear that this economic situation was to be imputed to more complex causes related to the technological changes in the film industry. The spread of multi-screen and multiplex structures has been promoted in recent years by new technologies (such as 3D shows) and successful Italian film productions.

Nowadays, digital cinema allows to realize more and more spectacular and fascinating products, changing the methods and models used by traditional cinema at the same time. Through the many applications of technology, films have become an integral part of the media system, instead of being autonomous and distinct. The impact of the new cinemas on the traditional ones has led the single-screen cinemas to get modernized and include author

Figure 12.

Figure 13.

programming and multiple activities: film clubs, drama, and dissemination. On the contrary, social multiplex attraction has been a real urban phenomenon, as well as demographic, as it has attracted commercial activities that derive economic advantages from this, creating a new centre in the city suburbs.

In the world of cinema the term "digital" is used with reference to the type of computer equipment adopted in certain stages of the life cycle of a film. If in the eighties the use of digital equipment had been introduced inside the post-production phase in order to intervene on the special effects, twenty years later it extended to many stages of film processing, from filming to soundtrack, from editing to special effects. The digital technology also spread in the context of data storage, but especially of distribution, implying the deletion of the print job on film favouring the electronic film transport. In recent years there has been more attention on development of the digital system, especially in support to small cinemas. In Italy the development of the digital system is growing, confirming its fourth position in the European scene after France, Britain and Germany. In particular, the Apulia Region, has allocated more than one million euro of Community funds to finance the transition to the digital system in about fifty selected cinemas by means of the "Digital Innovation and Technologies in cinemas" project.

Issues related to the conflict between multiplexes and smaller theatres are worth considering urban planning. The advent of multiplexes on the market has produced a dual effect on existing business: there have been the restructuring and modernization of the historic single-screen cinemas in urban centres, and the closure of smaller theatres unable to renew at the same time.

Currently historical centres tend to be used as qualified film locations in a more cultural offer and at the same time the most profitable; while multiplexes as places of solely commercial use. The survival of small and medium-sized theatres is due to their joining the Italian Federation of Art-house, whose "mission" is to promote high- quality cinema in the country and abroad, arrange meetings with authors in the halls, debate and disseminate informative

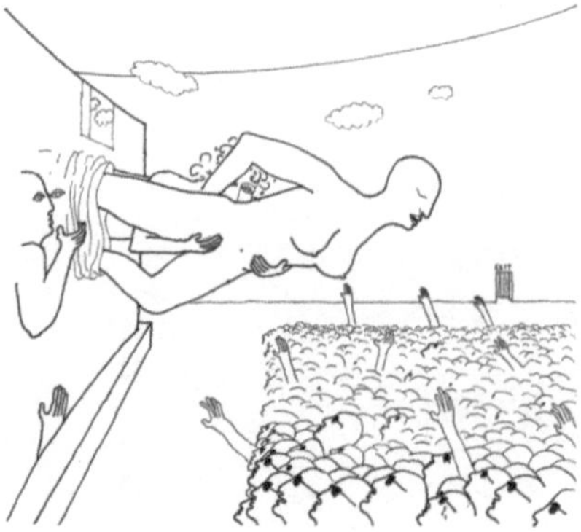

Figure 14.

Figure 15.

on the films shown. "Art-house" is classified as a high-quality genre. Also known as art-house cinema which literally means "experimentation", it refers to a collection of cinematographic works characterized by artistic, cultural or social contents, as well as experimental or unusual languages for a niche audience, who does not seek pure entertainment. It was born in France in the twenties and flourished in the sixties, it spread in all major European countries and the United States. Art- house cinema affected a minority group, although stable and constant, which was independent from fashion or trends of that moment. The context it belongs to is selective, but it tends to be followed by a wider audience because of the loyal relationship between them. In recent years, in order to renew small theatres located in old town centres, there have been projects and new strategies to promote high-quality show schedule and wider cultural offer. The activity of art-house cinema includes previews, meetings with directors, film festivals in their original language and educational projects.

Cinemas as new places in town, appeared first as a mixture with theatres and later went through a phase of extreme and unique identification. In the eighteenth century theatres gradually got closer to city centres without any alterations of the civitas; on the contrary, cinemas have been renewing in space, form and content. The high film quality of being so reproducible and saleable contributes substantially to determine the dynamics of the places intended for the cinema. The theatre, the cinema, the arena, the drive-in theatre, the multi-plex, the mega-plex are typical names and places for a kind of versatility. It is more than a century since the advent of cinema, but its evolution, especially from the architectural point of view, contains many elements that make it complex and fascinating.

The acceptation "cinema-theatre" identifies a limited space with screen and stage, meant for any kinds of theatrical performances. It is usually equipped with a scenic tower, where the

Figure 16.

Figure 17.

Figure 18.

equipment and the necessary mechanisms for the handling of the scenes are located. Between the two wars a prototype of a cinema with a stage started spreading. A mutual influence took place: old theatres were turned into cinemas, while new cinemas took the shape of theatres. However, the two functions had almost opposite characteristics: in theatres the audience tended to get close to the actors, the opposite occurred to films, because the image was too close to become distorted. So, despite showing no change in the location of seats, the most coveted seats in theatres were below the stage and in balconies, the best seats in cinemas were the ones in the back.

Cinema has inherited theatre simple formal passivity which is common to the two different entertainment places. In fact, in the twentieth century we witnessed the genesis of a controversial new building type, based on a formal, conceptual and decorative eclecticism. The analysis of that type of cinema showed that the changes, or rather the evolution of architectural spaces corresponded to a few key moments in the history of cinema. Firstly, the introduction of sound, that implied the beginning of acoustic studies and still represents an essential quality element. Secondly, the progressive autonomy of the show, corresponding to a longer duration of films, or the spread of length features films, that focused on the show more and more. The decorative element in the first films was a direct consequence of theatrical references. The first signs were generally obtained from pieces of fabrics provided with writings, put in the cinema halls. Later, with the advent of advertising, commercials became more and more organic, structured and coherent with the show, so that they appeared redundant, especially because of the large letters they were made of. The aggressive nature of advertising was tempered by the creation of new spaces for communication, which established visual order and dignity to the whole architectural structure. In the thirties a lot of cinemas started using neon signs, thus ushering in a new period of bright images, set in the busy city streets or in the surreal small towns, as an icon of change, dynamism and nature of contemporary cities. Soon even the facades of cinemas were turned into playbills and became the symbol of new communicative work for architects and managers. Some theatres became cinemas, and the first cinemas took the shape of theatres as well.

Cinemas got, in fact, the balconies of theatres that run on the side wings and the audience got closer to the actors; a useless feature for cinema because images, seen from aside, were distorted. There was no change in the arrangement of the seats: in theatres the best seats were below the stage and in balconies to better appreciate the actors' voice, in cinemas best seats for good view were the ones in the back. As to the show, there were differences in the

Figure 19.

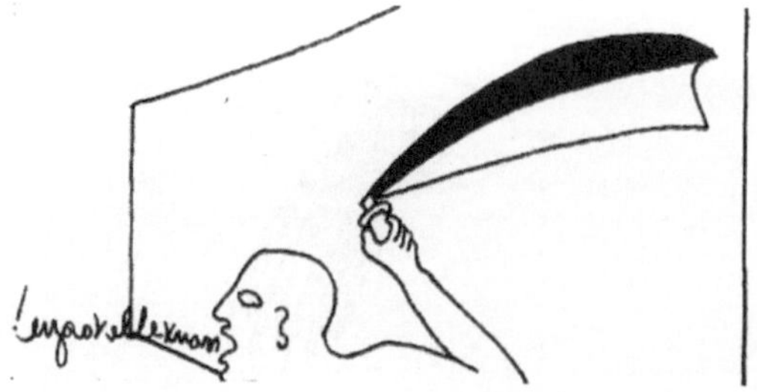

Figure 20.

ways of production: the times of filming were different from those of seeing a film, while in theatres those occurred at the same time. Later, with the search for a functional form and an architectural identity, gradually sumptuous and spectacular decorations disappeared. They were replaced, in time, by famous Italian and foreign artists' works: in 1948 Fontana made the ceramic friezes in the Harlequin Theatre in Milan. Cinema architecture was in turn influenced by the succession of architectural styles: from dec to liberty, from rationalism to postmodernism, up to advertising governed by the market economy. The evolutionary stages of cinemas have been marked by the commitment of various personalities of the Italian and foreign architectural culture, such as: Basile, Coppedè, Bazzani, Piacentini, Portaluppi, Mazzoni, Sabbatini, De Renzi, Free, Van de Velde, Wright, Poelzing, Taut, Mendelsohn, Scharoun, Garnier, Perret, Le Corbusier, Aalto, Asplund, Oud.

Cinemas became, therefore, places for shows which combined the spectacular nature of reality and the realism of fiction, placing itself at the same time as the embodiment of the new system of relations established by the metropolis, scanned by the alternation of working time and evasion. It is not therefore possible to trace the cinema history without taking into account its relationship with the world of architecture. And, equally, it is not possible to come to a full understanding of the multifaceted architectural events of the twentieth century, without taking into account the deep influence that cinema has had on the art of building. Since its birth, the film has established itself as a collective phenomenon, whose initial development was due to the strong expressive and emotional skills of the device, able to play real images as fantastic images which involve the viewer in a dark room. In time, the viewers' needs and unconscious have undergone modifications, assuming different characteristics from those of traditional film viewers. Its use has been collective until the fifties, when the television phenomenon and the advent of home entertainment lead to an increasingly individual type of domestic use of cinematographic works. In the following decades there has been, therefore, the television- cinema conflict, which has accompanied the crisis period of the film, especially in the eighties, when cinemas became the subjects of modernization, technology improved sound and image quality, and the film industry was renewed by producing high-cost films making them accessible to on the big screen only.

The architecture, the materials, but above all the interiors are made in order to reproduce typical imaginary urban environments. The buildings belonging to multiplexes have similar

Figure 21.

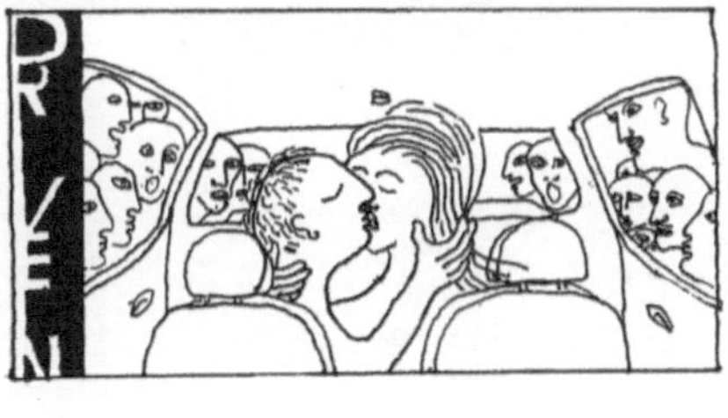

Figure 22.

architectural features to those used for shopping centres: the care for the forms is intended rather for the articulation of the interiors than the design of the exterior since the buildings are usually composed of a square body made of prefabricated elements and have no decorative details; the rare exceptions which reveal the presence of a greater architectural sophistication show the attempt to possess better identification. It is usual to combine the analysis of lighting and the curtain-wall in the main facade, to promote high visibility especially at night.

When automobiles became a means for the less well-off, there was also the growth in number of cinemas. In the immediate post-war years until the first half of the fifties, films started flourishing. A t first, cinemas were located in city centres because they represented the result of interventions on other buildings. With the advent of the automobile and the improvement of infrastructure conditions concerning mobility and public service, cinemas were located in more remote areas, too. A new type of cinema was built: drive-in theatre. Like an arena, it was located in an open and equipped space for showing films. The substantial difference concerned the organization of the audience that were parking in drive-in theatres. That type of film structure thus allowed to see films while sitting in their cars or motorbikes. Compared to arenas, the surface used for the stalls was much larger, so the size of a drive-in theatre was generally very large as well as the screen which allowed the sight from a distance. The metropolitan character of that type of structure implied a limited distribution in Italy. Indeed, in the second half of the twentieth century drive-in theatres spread in the United States, together with other facilities such as restaurants and coffee bars, allowing customers to entertain themselves by consuming their meal sitting in their cars. In Italy the first drive-in theatre opened in Casal Palocco in 1957, on the Roman coast. Like arenas, drive-in theatres became a mass phenomenon and embodied the spirit of road literature. However, limits due to seasons and the lack of choices compared to those provided by multiplexes contributed to reduce its spread.

Drive-in theatres inherited their spatial structures from arenas. "Arena" means a cinematographic structure located in an open space, specifically delimited, lacking a cover and equipped for showing films. In that area there is room for the audience; a screen, generally housed by a wall element or a metal structure covered by fabric; the projection box, which can be independent or connected to an indoor cinema which lies next to it. Arenas flourished in

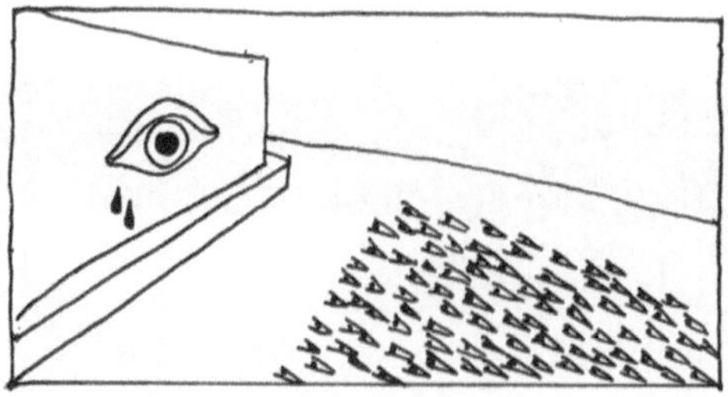

Figure 23.

Figure 24.

the period between the fifties and the sixties in connection with the economic development of the film industry when audiences became the protagonists of this phenomenon. At the same time arenas marked an era, imposing a new style of showing films in the summer. The limitations contained in a structure like an arena which can be used only seasonally, contributed to reduce the spread.

In the early nineties the frequent closure of old suburban cinemas led to the birth of cinema buildings together with shopping centres. In Italy and in foreign countries, the advent of multiplexes was not accompanied by a detailed and rational urban planning of the surroundings, which caused a great imbalance due to poor infrastructure. Although films gradually reached the awareness of an architecture style able to enclose their identities, the intermediate evolutionary stages of space research and adaptation always influenced and shaped the urban context in which they were located, penetrating all layers of society. This interference or indistinct introduction of cinema in urban and social city layers took place with surprise and curiosity; In fact, theatres represented the past in the collective unconscious, films embodied the idea of future, simply because made up of technological elements. Technological progress and the audience's interest lead to an increase in size of cinemas facilities until there appeared the first film with a separate projection box.

Beyond the architectural identity the power of cinema is to be found in the viewer, in his visionary ability and its temporary nature. Le Corbusier wrote "The real builder, the architect [...] can in fact create a magic box enclosing all that your heart desires. The moment the magic box appears it is a cube: with it comes everything that is needful to perform miracles: levitation, manipulation, distraction, and so on. The interior of the cube is empty but your inventive spirit will fill it with everything you dream of." With this statement, Le Corbusier thought of the symbiotic role between the architect and the viewer. However, the viewer's effort seems to be far more significant than the architect's. On the contrary, when the building takes on its true role these words can be understood: "The crowd is staring subjugated at the prodigious pavilion, they are getting on with no obstacles, as disoriented [...] they feel, without knowing precisely why, they are going to see something very exciting. This, for us, is true architecture." Those words represent Carlo Mollino's visionary idea, an architectural support to images and imagination.

Those places and their history, enclosing generations, have a magical atmosphere that transcends the historical space of the show, and become true urban fragments. Observing their decay, or worse, the profound alteration that often characterizes them, is, perhaps, an insignificant experience for those who are far from those places.

Figure 25.

Figure 26.

Figure 27.

Instead, those people who consider these "urban interiors" as transfigured shreds of an individual and collective memory at the same time cannot be unresponsive. That community is clearly nothing more than an elite group, a small circle of people devoted to the architectural form which contains the show. In fact, the viewing of a film cannot incur the interest in the architecture of the room, particularly when the viewing is replaced by simple consumption. The closure of the American Cinema in Rome, designed in 1954 by Angelo Di Castro, has raised clamour; it will be replaced by car parks and flats. Clamour is a sign of personal deep viewers' gratitude for those places, guardians of extraordinary visions and immaterial reminders of the architectural space.

Illustrations by Vincenzo D'Alba/Courtesy A.A.M. Architettura Arte Moderna Collezione Francesco Moschini, Gabriel Vaduva.

REFERENCE

Moschini F., Maggiore F., Ieva, V., D'Alba V., 2013. Territori del Cinema: Stanze, Luoghi, Paesaggi. Un Sistema per la Puglia. Letture e Interpretazioni. Roma: Gangemi.

Drawing (...) City (...) Body, Dwelling on Earth – Janeiro (Ed.)

The city skin – visions for a *tactile* identity of the space

F. Oliveira
CIAUD, Lisbon School of Architecture, Universidade Lisboa, Lisbon, Portugal

ABSTRACT: Solá-Morales proposed that the urban matter, which he refers to as the skin of the city, transmits us the collective energy of the city, carrying with it the complex meanings and cultural references that makes them a semantic matter, which carries the memory of places. The idea of a city skin, opens a concept for a tactile distinctiveness of the space.

There are fundamental rules for this to be structured, where functionalist issues impose themselves as distinctive of an ethical sense of architecture.

The moments of revelation, provided by a higher vision over the city's lower reality, display us designs, with an expression that transcends the mere dimension of the useful, sublimating some spaces to the plateau of Art.

The pavements often result, as links of symbolic connection to the City Ground, because they enshrine continuous structures of significant spaces, whose plastic expression makes possible the recognition of the territories of the walk, as mnemonic registers of the city. The subtleness of its variations, may cast the surfaces in such a distinctive way that, as a container of the most varied types of objects, the pavements became the space of the efficiency of walking and standing, which demonstrate, the relevance of walking the city as a poetic singularity, creating Architecture, by the liturgy of the stepped spaces.

Keywords: City Ground; Public space; Architecture of the city

1 TACTICAL IDENTITY OF THE PLACE

The philosophy of Merleau Ponty, makes the human body the centre of a world of experiences, and with this opens the field for the integration of the multiplicity of sensory experiences that characterize contemporaneity. The body assumes itself as a means of (in) corporation and synthesis with the surrounding environment: the world and the self, inform and continuously redefine each other. (Pallasmaa, 2006, p. 42) This extended phenomenology assumes as instrumental for the construction of an interpretation on the relation between phenomenology and architecture in general, and on the association between urban appropriation and the public space of the city in particular.

The aesthetic experience of the urban space, is assumed as a sensorial experience, where the qualities of space, matter, scale, are measured in equal parts by the eyes, ear, nose, tongue, skeleton and muscles. (Pallasmaa, 2006)

In the (bio)logic of cells, it is by their multiplication, according to precise patterns of variation and arrangement that one reaches the concretization of these in organs and distinct functions, proper to an organism.

The city, as a system, grows and organizes itself nourished by a bipolar intelligence that is structured between a formalization, typified by distinguishing factors and the deterministic specification that allows the concretization of its autonomous urbanity, in unique urban contexts. In a similar logic, if the first we can call spatial macrosingularities, the second ones will be resisted as local microsingularities. (Vigliazzi, 2003)[1] Thus, a parallel is established with the idea of skin as the primordial organ of the body and simultaneously of fingerprint as a distinguishing factor and sensorial individuality.

> "I experience myself in the city, and the city exists through my embodied experience". (Pallasmaa, 2006, p. 11).

In similar terms, the multiple components of the urban space match to the unitary elements of the constitution of identity. They are equivalent to the cells of an organism that, grouped in different logics, define and compose various organs and correspond to public space systems, of a being that constitutes a City.[2]

Solá-Morales suggested that the urban matter, which he refers to as the skin of the city,[3] transmits us the collective energy of the city, carrying with it the complex meanings and cultural references that makes them a semantic matter, which carries the memory of places. (Solá Morales, 2008, p. 3).

In another dimension, the (re)definition of the memory of the city concept, as a critical register of experiences, presumes, the capture, the retention, and the forgetfulness (Louro, 2016). In other words, in the process of perception, a selective reading of records and omissions is emphasized. An evaluative judgment in which the city is revealed as palimpsest, whose conformation logic is born from an experienced vision and touched by the body.

The question of inhabiting and the appropriation of space as a model of affinity with the surrounding context is established in a unique way within the framework of everyday life. In this way, the appropriation of the space that is inhabited is assumed as an intention to modify it and adapt it to the contingency of a subject that participates in the same space. Here, the Heideggerian structure, between building, dwelling and thinking, assuming the building,[4] as a means to the end of dwelling, and defining dwelling as the fundamental trait of Being (Sein). (Heidegger, 1996)

1.1 *Topogénetics and the sublimation of the place*

The search for the deep sense of place, starting from a topogenetic[5] approach, is a well-founded and rigorous strategy of creation, where an architecture, and by extension a city, are strongly related to the context, and might restore the principles of contemporary coexistence, structuring the balance and harmony of the urban territories. Contrary to the forces of rampant and deterritorialized globalization, we find here a sense of respect for a dialoguing reality with the preexisting authenticity. This orientation towards modernity, intends to reconcile, the freedom of creation with the flows of continuity and tradition that surround the creator. Allowing the development of dialogically structured attitudes, as suggested by Mikhail Bakhtin in his philosophy.

---

1. Vigliazzi, defines the fingerprint as being the representation of the epidermis of a finger. The fingerprint formed by a set of lines that produce local macrosingularities and has the characteristic of being unique in each individual and does not change with the passage of time. Microsingularities also known as minutiae (minutiae) are determined by the bifurcations or terminations of the ridge lines. Most algorithms use minutiae for fingerprint identification.
2. Reference to the relationship between the anthropometric view of the architecture based on references to Vitruvian Man, the Fibonacci series and the Modulor and the interpretative review of biometrics as a strategy for reading and interpreting urban space.
3. Manuel Solá-Morales, assumes the skin of the city as an entity that is made of constructions, textures and contrasts, being by the contact between the body and the physical matter that the urban experience arises. (Solá Morales, 2008). The reference to the meaning of the word build in German: *bauen*, which not only means build but also *cultivate*.
4. The reference to the meaning of the word build in German: *bauen*, which not only means build but also *cultivate*.
5. The word/concept that derives from the concept of the Topogénese of Muntañola, with the idea of a genetics of places, associated with the idea of Greek genus, of being born and of the way in which the characteristics of living places are transmitted.

Codes and laws that articulate language as a logic of being and as a paradigm of its identity, link the place as an external logic, as a relation inscribed in a territory. Architectural poetry is not limited to creating an envelope for the actions of man; it tries to recreate the actions and desires of those who inhabit them, thus strengthening the bonds between that and their users.

The act of conceiving architecture works as a mover of reality, through which, in addition to being able to make explicit, it also shapes its future, impelling the generating force of other real, imagined and constructed.

Every place built necessarily implies a logic. There are fundamental rules for this to be structured; there are practical questions that require basic foundations for architectural creation, where functionalist issues impose themselves as distinctive of an ethical sense of architecture. It is through the use of architecture that the impression of time in space is allowed, wandering in these, to the flavour of the variants of daily life, transforming buildings and cities into an indelible idea of the reality of wanting to exist.

The understanding of the ethical dimension of places emerges from respect for the preexistence and the involvement, to allow them to coexist in the sense of mutual esteem. Affirming ties of affection and memory essential for the cultural identity of societies. The place is aesthetic, because, as the stage for existence, the architectural space is placed between the subjects and the objects, celebrating different spatialities according to the use and its users. But there are ethics in this aesthetic, where one tries to understand that, the aesthetic questions adjust according to the ethical perspectives of organization, predetermining, what can be called, normality parameters of an essential architecture.

> "In the relationship of the body with space, the building is not the only architectural place. The whole context is architecture." (Pierre Pellegrino).[6]

Architecture, articulating the physical and the geometric, formalizes systems of reference in the dialogism of spatial articulations, as Muntañola explains to us through the teachings of Mikhail Bakhtin.

> "The triple architecture of time and the three social dialogic dimensions are the two basic pillars of the chains of life. Albert Einstein was probably not aware of the astounding power of his own theories. So we are at the beginning of a relativistic theory of the universal architectural values of ours, in which diversity is the condition of universality and universality a condition of diversity. However, we treat architects with social, mental and natural powers, as our brains do, and not only with natural or technical powers, as our machines do. An architecture of human dialogic responsibility is my dream, and I believe it has also been the origin of all the dreams of human beings throughout history." (Muntañola, 2000, p. 7).

Places, in upholding, the complexity of things, formalize themselves as foundations for a relationship between subject and history. In this sense, the City Ground assumes itself as a refined synthesis of experiences and registers, marks of subtle intensity that define times and memories in a sublimation of the place. (Oliveira, 2017).

Observing only the hardware of the city is entirely insufficient and reducer of its real identity. The reading that is suggested introduces, through the emergence of the concept of the City Ground, the promotion of semantics of urban space, which complements the physical view of the context. Resuming Josep Muntañola, the context accommodates the physical and material part of the City (the hardware) and reveals itself as the generator of dialogic interactions, where text, the content (software) is installed exposing intentionality and promoting identity.

In architectural creation, the stimulus that presides over the creative act encompasses in its purpose the function. Being the sense of usefulness, determinant for a characterization

---

6. Fragment of the Prologue of Pierre Pellegrino to the first edition in French of the book: Josep Muntañola Thornberg, Topogénesis – Fundamentos de una Nueva Arquitectura, Barcelona, Edicions UPC, 2000 (1ª edição 1996), p. 12.

of the architectural object[7]. (Rodrigues, et al., s.d.) The concept of habitability constitutes, therefore, as the main motor of the architecture, defined by a strong continuity between the building and the dwelling. (Heidegger, 1996).

In this context, the relationship with urban space is structured, where the premise of usefulness is established in the Aristotelian relation: city, utility, happiness. That is, useful is assumed as a resource to provide happiness. In this way, the useful is determined by the weight of the urban and its collective wearer, mediated for the ultimate purpose of happiness. Happiness gives the concept, the notion of meaning itself, and in this way, in the City, objectifying the useful establishes itself as an objectification of the beautiful.[8] (Lalande, s.d., p. 144)

1.2 *The architecture of the city, the public space as Axiom*

The allusion to the architecture of the City implements the concept of Aldo Rossi, that derives from the understanding of the city like an architectonic act. For Rossi, Architecture is the structure that mediates the subject and the territory, as a materialization of the spirit, as a response to the ethical, aesthetic and ontological needs of the being.[9] (Rossi, 1976.) It appeals to the sense of usefulness and its plastic and spatial value. The architecture of urban space is thus understood as the synthesis of a process of creation of the city, as alluded by Nuno Portas, (Portas, 2007) in which the City Ground materializes its existence, as a metaphor for its process of (con)formation. (Oliveira, 2017).

> "Street pavements represents the low-water mark of the urban environment." (Rudofsky, 1964, p. 265).

> "Public spaces must be viewed in three dimensions, as volumes carved out of the solid of the built fabric." (Kostof, 1999, p. 138).

This often-overlooked notion, that the space between forms is also a form. This fundamental conception allows us to gain full awareness that, there are no separate forms and that there is always, as Fernando Távora revives us, a relationship between the forms we see occupying space, which we know to constitute the form (negative or cast) of the apparent forms. (Távora, 1982, p. 24).

The public spaces, as a collective heritage, must respond to the transformations and dynamics of today's city dealing with the ubiquity of memory and the response to new urban trends, in a continuum of complementarity between generations. Due to the collapse of the environment that allows their interaction, the dissolution of the link between the subject and the city forms a

7. The sense of useful is assumed to be something that serves life or more specifically happiness. In the background that serves a practical purpose. What has no value in itself, but above all as a means to achieve a specific end, from any point of view. The useful is thus the humanization of the created work, understood as a representation of purpose, in a precise, temporal and immediate way. In the context of architecture, it is assumed as a fundamental assumption. In the relationship between artistic and architectural creation, the useful and the beautiful, are assumed as complementary concepts. In the background the search for beauty in artistic creation, is defined by a search of useful in architectural creation and consequently urban. In architecture, the usefulness is thus insinuated as a fundamental stimulus of creation, as fundamental in the programmatic composition and its formal resolution. (Lalande, s.d.) e (Rodrigues, et al., s.d.)

8. On the concept of beauty: One of the three fundamental normative concepts to which the judgments of appreciation can be linked: it is in this sense that it is defined by Kant: "What pleases universally without concept."

9. For Aldo Rossi, Architecture is the key to a correct interpretation of the city as a structure in space. The "Urban Science", as Aldo Rossi designates Urbanism, will be the study of the city as Architecture. The controversial attitude of Aldo Rossi is based on understanding Architecture as an autonomous value, at least in a certain dimension. To revalue the singular work and the monument as a fundamental element of the history of the city and of the collective memory, in the face of the pragmatic interpretations that make the architecture a trivial point in the structure of the city. For Aldo Rossi, Architecture continues to have a qualitative dimension that cannot be mortgaged and that is essential to understand the collective realization of the man who is the city.

new paradigm in the discussion of the problematic of the contemporary urban condition, configuring the notions of public space as an axiom,[10] versus the public space as a consequence.[11]

The ground that we tread upon, and which, with our bodies we alter, is sometimes absolutely anonymous on the speech of the City. Magnificent examples of design and creativity are lost in the glance of the ordinary user of the city. The, almost magical, suggestion offered by the relocation of the eyes to a higher view, unveils an extraordinary vision over the city of men, so well revealed by the symbolic gaze of the angels Damiel and Cassiel when, in the Wim Wenders film, they watch over those who, in the anonymity of the ground, endure in silence. (Der Himmel über Berlin, 1987).

The moments of revelation, provided by a higher vision over the city's lower reality, display us designs and intentions, often hidden, in the enigmatic work of artistic revelation, with an expression that transcends the mere dimension of the useful, sublimating some spaces to the plateau of Art.

The City is one thing that exists, and that constitutes itself beyond the strictly architectural. However, it is subordinated to reading and composition systems that use the interpretive tools of architecture concerning the primary processes of civilization, structuring the relation of man with the universe. The assembly of the imaginary of places that are beyond the visible and immediately perceptible reveals to us, eventually, the intelligence of these places and of those who have shaped them. The imaginary, as the fruit of personal experience, in complement with the symbolic as a reflexive, collective and cultural phenomenon, implies a practice, an experience, a semantic memory of the place.

The absence of this memory, of this imaginary, compels us to the alienation, revealed by Juhani Pallasmaa, when he establishes the counterpoint between the city of participation, as a place of memory, experience, and sensorial commitment, as opposed to sensory deprivation of the city of alienation. (Pallasmaa, 2006, p. 45).

The instrumental properties of the architecture revealed as an operative, and real actions are supported by the physical components of the buildings, such as materials, colors, textures, and shapes, which allow organizing the way of perceiving the city. Its skin, the surface of Its Ground is, in a way, defined by its instrumental properties. However, a vast set of relational properties are assumed as a general and abstract condition, inducing an urban architecture. These reveal the creative abilities, in the intentional combination of proportions and relative relations, where, by means of geometric models, they are allowed to conjugate the hierarchies and spatial arrangements.

1.3 *For a tactile appropriation of the city*

From the structure of the macrosingularities, lays the assumption of the need to establish the minutiae as marks of the tangible contact with the city, that is, find the meaning of the words contained in the key phrases of the urban discourse.

> "The taste of the apple ... lies in the contact of the fruit with the palate, not in the fruit itself; in a similar way ... poetry lies in the meeting of poem and reader, not in the lines of symbols printed on the pages of a book. What is essential is the aesthetic act, the thrill, the almost physical emotion that comes with each reading." (Borges, 1972, p. 272).

The feeling, alluded to by Jorge Luis Borges, in the "thrill" that provokes the mutual changes incited by the contact between the subject and the environment, is responsible for the altering of the empirical relationship between the subject and the city space. This fundamental action, allow us to better understand the response indorsed by the caressing of space with the voracity of a glimpse of the eyes, and so often crystallized in the instant of a photograph or in the delight of a drawing.

---

10. This axiomatic sense derives from the notion of public space as a postulate of the construction of a city, as something unquestionable, unequivocal and active, as a primordial intention.

11. In this case, public space is understood as a product that derives from a process of urbanization, which insists as mere spatial occurrence and not as spatial intention.

The imprisonment of light and shadow, at the confrontation with the materiality that supports them, raises canvas filled with memories and incitements, which guide our senses through a delight of emotional journeys in which is offered the opportunity of touching the space, once sensed, in a dreamer's "tactile fantasy". Thus, revealing the idea of an architecture of human responsibility, absolutely dialogical, dreamed by Muntañola.

The subtle details that emerge from the contact marks between the subject and the city establishes the true meaning of the concept of microsingularity as a distinctive component of a place. These city fragments are like body projections, which through the movement across space become architecture, and these become architectural because of the ability that they have to retain our thoughts, in the sense in which they enable the encounter between the dream and the dreamer.

The light is transmuted into an argument, in the sense in which it models the frontier territories between seeing and feeling. Here where the presence of the luminous and the shadowy suggests the existence of parallel worlds, the penumbra zones carry the space of the mystery, proper of the imagination, giving shape and energy to the enlightened space. In the words of Alberto Campo Baeza, light gives reason to time, constructing time itself, with architecture chasing light, shaping and drilling planes so that the light surpasses them and materializes in space through the play of tonalities between light and shadow. (Baeza, 2013, p. 67).

It's in the details, in the minutiae of its microsingularities that the city reveals itself as Architecture. The instrumental and relational properties that it carries emerge from materiality that is fused by higher ordinations, expressing the intentionality of a certain sense of order. The careful observation of the signs engraved on the skin of a city, allows us to take an extraordinary journey into the core of the imaginary structure of places. In which, the presence of time, and its passage, witnesses the creation of authenticity, where the "tactile memory" allows us to capture the tangible sensations from a distance, as it is suggested by Pallasmaa. "With the vision, we touch the sun and the stars." (Pallasmaa, 2006, p. 40).

Feeling the Ground of a city is the most inevitable of the urban experiences. This tactile appropriation is the one that defines us as a body, with mass, weight, with material existence.

The expression of plasticity, the sound it reflects, the freshness or the sunshine that it emanates, the light it exudes, the moisture it absorbs or repels, the cleanliness it demands or facilitates, or the respect that inspires or declines, are qualities of which we may have, or have not, consciousness but we cannot be surprised when we are walking, touching the Ground of the cities that we step.

1.4 *The path as poetic singularity, the urban pilgrimage and the liturgy of the stepped space*

Architecture creates spaces from which we can enter and exit. Being able to enter means that the experience about us is such that, paradoxically, it's they who end up entering us. Architecture, as the generator of spaces, does not have a beginning or an end. This is what enshrines the uniqueness of the architectural poetics and which Le Corbusier put in evidence through his concept of "Promenade architecturale".[12]

Architecture can be felt by the ability to make space exist, to modify it, to make it distinctive. It is to its uniqueness that architecture owes what makes it so difficult to define.

A city is not an object, it is a structure. In this structure, architecture engenders the harmonization of places. A place opens on another place, and the displacement is the celebration of

12. Promenade architecturale, while valuing the route as a conceptual strategy, to order both internally and externally the architecture, punctuating the experience of fruition of the architectural object with constant surprises. The concept reveals itself through a set of material properties, consciously worked with the purpose of realizing the idea of variation of the path, forcing the experience of the architectural object in different positions and points of view and constantly varying the relation between the object and that object which frui. Forms on the light, inside and outside, above and below. Inside, we enter and walk, we watch as we walk, and the forms explain and develop, combine. Outside, we approach, we see, we care, we enjoy, we travel around, we discover ... We do not stop receiving successive emotions.

the places. The displacement must be understood, as a crucial event. It belongs to the existence. To exist is to move, existence is displacement. Displacement is our essential dispersion.

The experience of the architectural place is strongly committed to the issues of human existence in time and in space, both of which structure the experience of the integration of the individual in the world, reinforcing the sense of Being. The Experimentation of space, walking in the city, builds the sense of belonging to the place and its memory. (Roca, et al., s.d., p. 301)

Pilgrimage is the expression of man's nostalgia for transcendence, as a fundamental experience of spiritual appropriation. It is the evocation of the memory and the celebration of intimate values, between that pilgrim and the symbolic of the liturgy of the travelled space.

> "The pilgrimage goes in freedom, like a walk of friends in beloved land, long not seen, but where nothing will be lost of its essential: what has soul and what has beauty." (Araújo, 1992, p. 9).

The idea of crossing a space is born from the primordial need for the movement to find food and information essential for the survival of the human species. However, once these primary needs were assured, the act of walking became a symbolic action that allowed humankind to inhabit the world. By transforming the meanings of the crossed space, the path becomes an aesthetic action. And by this, building the order on whose bases the architecture of the objects that inhabit it is built. Walking is an Art that encloses in essence, architecture, and landscape.

> "The city discovered by the vagabonds of artists is a liquid city, an amniotic fluid where other spaces spontaneously form, an urban archipelago to navigate by drifting: a city in which living spaces are like islands of the immense ocean formed by the spaces of the walk." (Careri, 2002, p. 21).

Walking can be understood as a (dis)placement, meaning that there are always places, but also spaces other than places, spaces between places, and therefore places float, do not remain stable. Displacement is a critical act and walking is thus an aesthetic instrument.

The pavements often result, as links of symbolic connection to the City Ground, because they enshrine, by their design and materiality, continuous structures of significant spaces, whose plastic expression makes possible the recognition of the territories of the walk, as mnemonic registers of the city. (Oliveira, 2017).

> "The city appears to us as a whole in which no desire is lost and of which we are part, and with it enjoys all that we do not enjoy, we can only inhabit this desire and satisfy ourselves with it." (Calvino, 1990, p. 16).

The spontaneous casuistry of the city's spatial formations can and should be converted into intentional project activity. The horizontal structures of the city, in their complexity and comprehensiveness, play a decisive role in the construction of the Ground places, and by their design and materialization, they become a receptacle of our material existence and our creative identity.

The conscience that little is known about the importance of these structures of treading refers us to the relevance of the emergence of this surface of contact with the Being, as an opportunity for innovation. This desire to inhabit the city, allows the encounter with a City Ground, eternally postponed, as the place for the happiness idealized by Aristotle.

The pavements, the skin of the City Ground, emerges, in this context, as an instrumental fusion between the ethics of doing and an aesthetic of feeling. The variants of its existence allow the faculty to assert themselves as protagonists in the act of claiming the city as an inhabited structure, subtly designed by all those who have stepped on it and who still dream of touching on it.

The pavements accuse the tangibility of the Ground; they are part of its ROM memory[13], the one that is recorded in the stones, engraved by all those who step on them.

---

13. ROM stands for "Read Only Memory". It is a memory that facilitates the preservation of information that can be read but can not be destroyed.

The materiality and subtleness of its variations, may cast the surfaces in such a distinctive way that, as a container of the most varied types of objects, the pavements became the space of the efficiency of walking and standing, which demonstrate, the relevance of walking the city as a poetic singularity, creating Architecture, by the liturgy of the stepped spaces.

REFERENCES

Araújo, N. 1992. Peregrinações em Lisboa (Vol. 1). Lisboa: Vega.

Baeza, A.C. 2013. A ideia Construida. quinta edição ed. Lisboa: Caleidoscópio.

Borges, J.L. 1972. Selected Poems 1923–1967. New York: Delacorte.

Calvino, I. 1990. As Cidades Invisíveis. Lisboa: Editorial Teorema.

Careri, F. 2002. El andar como Prática Estética - Walking as an Esthetic Practice. Barcelona: Editorial Gustavo Gili, S.A.

Der Himmel über Berlin. 1987. [Movie] Directed by Wim Wenders. Germany/France: Road Movies and Argos Films.

Heidegger, M. 1996. Bâtir, Habiter, Penser. Em: Essais et Conférences. Paris: Editions Gallimard.

Kostof, S. 1999. The City Assembled The Elements of Urban Form Through History. London: Thames and Hudson.

Lalande, André. s.d. Vocabulário Técnico e Crítico da Filosofia, Vol. 1. Porto: Rés-Editora.

Louro, M. 2016. Memória da Cidade Destruida: Lisboa/Chiado – Berlim – Sarajevo. Primeira Edição ed. Lisboa: Caleidoscópio.

Muntañola, J. 2000. Topogénesis - Fundamentos de una Nueva Arquitectura. segunda edição (1ª edição 1996) ed. Barcelona: Edicions UPC.

Oliveira, F. 2017. A cidade como Chão de palavras pisadas. A arquitetura dos pavimentos e a identidade dos lugares da lusofonia.. Em: Academia de Escolas de Arquitetura e Urbanismo de Língua Portuguesa (eds.), A Língua que Habitamos – VOLUME 1 Condições de uma Cidade com Futuro: Cidades Antigas, Novos Espaços. Lisboa: AEAULP, pp. 95–100.

Oliveira, F. 2017. Texturas, Pavimentos e Chão – Um Chão de (autenti)cidade, pavimentos e outros complementos.. Em: R. Duarte, M. Louro, F. Oliveira & M. Gamito, edits. AutentiCIDADE – Considerações sobre a condição urbana contemporânea. Chiado Editora, Lisboa, 2017, pp. 105–133. Lisboa: Chiado Editora, pp. 105–133.

Pallasmaa, J. 2006. Los Ojos de la Piel – La arquitectura e los sentidos. Barcelona: Editorial Gustavo Gili, SL.

Ponty, M. 1992. O Olho e o Espírito. Lisboa: Vega.

Portas, N. 2007. A Cidade como Arquitectura – apontamentos de método e crítica. 1ª edição 1969 ed. Lisboa: Livros Horizonte.

Roca, Estanislau., Aquilué, I. & Gomes, R., s.d. The city as memory, Interpretations and representations. Em: E. Roca, I. Aquilué & R. Gomes, edits. Walking the city: Barcelona as a urban experience. Barcelona: Publicacions i Edicions de la Universitat de Barcelona, pp. 301–311.

Rodrigues, Maria João Madeira, Sousa, P. F. d. & Bonifácio, H. M. P. s.d. Vocabulário Técnico e Crítico da Arquitectura. s.l.: Quimera.

Rossi, Aldo. 1976. La Arquitectura de la Ciudad. Barcelona: Editorial Gustavo Gili S.A.

Rudofsky, B. 1964. Streets for People – A Primer for Americans. Garden City-New York: Doubleday & Company, Inc.

Solá Morales, M. 2008. De Cosas Urbanas. Barcelona: Gustavo Gilli.

Távora, F. 1982. Da Organização do Espaço. Porto: Edições do Curso de Arquitectura da Escola Superior de Belas Artes do Porto.

Vigil, Gustavo M. 2000. Diseño Urbano – Teoria y Metodo. Mexico: Alfaomega Grupo Editor.

Vigliazzi, D. 2003. Biometrias – Medidas de Segurança, Visual Books, 2003. s.l.: Visual Books.

Drawing (...) City (...) Body, Dwelling on Earth – Janeiro (Ed.)
© 2019 Taylor & Francis Group, London, ISBN 978-0-367-02602-8

The classical composition in the social spaces of housing in Recife, Brazil

G. Carvalho
CIAUD, Lisbon School of Architecture, Universidade Lisboa, Lisbon, Portugal

I. Bechmann & M. Figueiredo
Damas Da Instrução Cristã Faculty of Architecture, Recife, Brazil

ABSTRACT: Given the most diverse occurrences of classicism throughout the history of architecture, this article intends to analyze to what extent the classical compositional principles are present in some moments of the history of housing in Recife, Brazil. In order to do so, we first discuss the relations between aesthetics and housing architecture, considering the issues of perspective design, image and classicism in architecture, adopting as classical compositional principles those elaborated primarily by Vitruvius. Four representative examples of the history of housing in this city are analyzed, since these principles become part of the architectural design procedures adopted in the country, XIX, with the installation of the Portuguese Court in Brazil, and the creation of the National Academy of Fine Arts. The examples analyzed are a classicist dwelling, an "art-deco", a modern and a contemporary one, particularizing the social spaces of these dwellings. The analysis extended from the composition in plan, as well as the main façade, entering the interior composition of social spaces, and concludes by demonstrating that the Vitruvian compositional principles were identified in all of them, contributing to a wider investigation, which seeks out aesthetic design references which have spanned centuries, and are presumably still recurrent in today's housing projects.

Keywords: Classicism, Interiors, Housing, Composition, Aesthetics

1 INTRODUCTION

In the analysis of housing architecture projects in Recife, it is nowadays perceived that social spaces continue to be more valued from the point of view of their plasticity, design, coating materials and compositional elements, in relation to other housing spaces. Independently of the models that the housing type assumes in the city of Recife, whether in individual housing projects, apartment buildings, housing interior projects, and in closed condominiums, one perceives an intention to value social spaces, interfaces between the public and the private, in view of an intended symbolic image of these spaces to its observers.

It has also been verified that the value of these spaces and the concern with their appearance are independent of the socio-economic aspects of their users and of the value of the property, in a country where a large part of its population lives in spaces not designed by professionals, and where the dimension aesthetics, considering the classic triad of Vitruvius referenced for architecture, solidity, functionality and beauty, is questionable in the face of the most pressing needs of shelter.

From this, two questions arise here. The first one concerns the permanence of a habit of valuing these spaces, dating back to the 19th century, when the city dwellings began to invest in their social spaces, in keeping with the new habits of society acquired after European cultural influences, notably the French. The questions of appearance, in architecture, that is, questions related to the image, will assume a prominent character in the architecture projects observed since then.

Another question, directly related to this first one, is the fact that, for the appreciation of these spaces, and to compose them architecturally, the classical compositional parameters still seem to prevail. This is verified both in the design of the building and in the design of its interiors. The classical reference and its guidelines are the ideal of beauty that has prevailed since the installation of the Portuguese Court in Brazil, and the subsequent adoption of classical compositional principles in the teaching of the National Academy of Fine Arts, with its repercussions on the professional practices of the professionals since then, and in the adoption of the repertoire in fashion from the craftsmen, craftsmen and society in general, in free form, but also imposed by the Municipal Postures of the main cities.

2 IMAGE AND ARCHITECTURE

The design, which precedes the project, which materializes the idea of the professional architect, is identified by Vasari as its foundation (Chaymovich, in Tejo, 2012:20). As far as architecture is concerned, the advent of perspective technique will be a fundamental achievement in the representation of the space devised by the architect, as well as in the conception of the project. However, perspective representation is relative to views set out in a plan, which eliminates the other views, also possible to represent a "possible world." These other visions, which would lead to a more accurate understanding of reality, lead to an incessant game of questioning about the represented space, which is therefore imprisoned in a single, chosen by the artist/architect.

This question points to the problematic of composition and classicist representation in architecture, which is valued, it seems to the present day, and consequently in the social spaces of housing, object of this investigation. The scene is privileged, the angle of view, and usually the background is defined from the main access of the dwelling, in the sense of who enters it, addressing its social spaces.

This character of scenery will be accentuated from the classical compositional principles, transforming the spaces into true "finished scenes". However, perspective does not represent space in its entirety, and does not encompass all the dimensions that an architectural space can elicit in man. The aesthetics of architectural space tends to compose it only from the point of view of design, of its constituent elements, privileging the physical-aesthetic dimension of architecture (Janeiro, 2010:424), to the detriment of its much more complex essence. This is because "o espaço simulado pelo desenho é um espaço só visual"[1] (Janeiro, 2010:417). According to him, the drawing is a "becoming", an initial stage. It is no longer the idea, since it has already materialized in drawing, but it is not yet architecture (architectural object). However, the drawing "seduces," and causes the subject to see himself within this space, and to glimpse a drawn reality. The tendency is for the subject to be transported to a place where everything is defined, organized, aesthetically beautiful.

Certainly, from the Italian Renaissance, the classic reference will be followed in the composition of the housing architecture projects, and not only. It will enter as a reference in all the architectural typologies, and throughout the following centuries by the most diverse architectural styles. And what is perceived, when analyzing the projects of housing architecture in Recife, is that their compositions have an aesthetic reference that remains throughout the centuries, and that is based on the classical compositional parameters.

The second point of interest of this analysis, which concerns a habit that dates back to the nineteenth century, and which still remains, is that of valuing social spaces in the houses of Recife, Brazil, in view of the aesthetics of these spaces, since they will be the spaces used by

1. "Simulated space for drawing is only a visual space" – free translation of authors.

the visitors. Spaces that value the appearance, intending to represent the socioeconomic reality of the owners. These environments are worked within a scenario context, where classical compositional principles prevail, and where greater financial resources are invested, in a clear purpose of demonstrating economic power and social status.

Concerns about appearance are identified in societies where there are large differences in social classes and demonstrations of power among citizens. This is identified in Recife, Brazil, but also in ancient Rome, and theorized by Vitruvius in Book 5 of his Treaty. Vitruvius discusses "common and private parts of houses", indicating a differentiation of rooms that are destined to "homeowners" or "strangers", probably the origin of a practice that goes back to the present, to differentiate intimate rooms from social rooms in a housing. This book also mentions a "adequação da casa à função e estatuto do proprietário"[2], where it indicates the need of the economically rich people to own a house of greater aesthetic representation and also that it contains a greater number of rooms, "condição indispensável para que exerçam a contento as funções sociais e cívicas inerentes a suas posições."[3] (Vitruvius, 2007:312).

Therefore, in the housing architecture projects in Recife, the aesthetic dimension, the emphasis given to the visual, has the purpose of transmitting an image to its observers, and the image that predominates within this investigation is supposedly the classic one. Gustavo Rocha Peixoto, in his work "Reflexos das Luzes na Terra do Sol"[4] discusses the values of the neoclassicism that it contributes here, like "... modelo de civilização, em oposição à barbárie reconhecida pelos lusos recém-chegados como medieval e mourisca."[5] (Peixoto, 2000:281).

In the eighteenth century, classical compositional rigidity will be the guiding element of palatial interiors and rich bourgeois houses of Portuguese society of the time, which in turn will bring these influences to Brazil in the first half of the nineteenth century. We will also see that classicism has been identified in the housing architecture both in its interiors and in the building itself, and one of the reasons is that, in the history of architecture, the composition of the interiors will follow in many cases the classicism determined by the formal nature of the construction, that is, it is verified that the projects of internal composition were as a result of the classical composition of the exterior.

So the perception that contemporary housing projects in Recife remain with their classical composition is due primarily to the appreciation of classical aesthetics by society from the 19th century, and secondly to a pretension of the more affluent classes to pass an image of power and social status, but also of erudition, "good taste" and modernity, summarized, symbolically, in this aesthetic.

3 VITRUVIAN COMPOSITIONS PRINCIPLES, CLASSICISM AND ARCHITECTURE

Discussing classicism in architecture, a theme so widely reflected and documented, leads us first to consider that classicism is a guideline for us to achieve a harmonious composition in architecture, an "intelligible harmony", always sought by the architects. Perhaps that is why compositional patterns dating back to classical antiquity are still present today: a formula for harmony and beauty. Here, we will take as a reference of classical compositional principles those elaborated by Vitruvius in his Treaty of Architecture, discovered in 1414, to Renato Brolezzi, "... elevado à natural condição de autoridade suprema, suas consequências foram avassaladoras... em diferentes escalas e contextos"[6] (Vitruvius, 2007:26). From it derives the

2. "Appropriateness of the house to the function and status of the owner" – free translation of authors.
3. "An indispensable condition for them to exercise to the satisfaction of the social and civic functions inherent in their positions." – free translation of authors.
4. "Reflexes of the Lights in the Land of the Sun" – free translation of authors.
5. "... model of civilization, as opposed to the barbarism recognized by the new lusos as medieval and Moorish." – free translation of authors.
6. "... elevated to the natural condition of supreme authority, its consequences were overwhelming... in different scales and contexts" – free translation of authors.

triad firmitas-utilitas-venustas, translated by Justino Maciel as solidity, functionality and beauty. We are concerned with the aesthetic dimension, with the venustas, since what guides our interest is precisely an investigation about the compositional principles that govern the conception of housing projects, and considering its fundamental importance for classicism. We identify in chapter 2 of Book 1, six Vitruvian compositional principles, namely, ordenação, disposição, euritmia, comensurabilidade, decoro and distribuição.[7] (Vitruvius, 2007:74). These parameters merge and interpenetrate, since the differences between them are very tenuous.

The concept of ordenação, or "ordering action," is a proportion of the parts that make up space separately, but also in their entirety. The concept of disposição refers to the proper placement of the elements and the aesthetic effect that results in space. The concept of euritmia is proportionality between all parts of the work, height, width and length, also translated as harmony. The concept of comensurabilidade, translated as symmetry, that is, a correlation, a system of measures between the parts of a work that generates a final balance, different from what is understood today, as a "mirroring", a reductionism of the concept. The concept of decoro, a convenience of using the elements of architecture that is use "what is good" which suits each particular case. And finally, the concept of distribution, that is, to appropriately allocate the means available in a certain place for the execution of the architecture, instead of looking elsewhere, burdening the work and making it difficult.

The Treaty of Architecture of Vitruvius has been influencing Western architecture ever since. Its principles come to the Modern Movement of the early twentieth century, making its protagonists, even denying the figurative adornments and historicist references, have been based on significant works in the housing architecture, such as the Muller House of Adolf Loos, and the Villas by Le Corbusier. As an example, we have Colin Rowe's comparison between the Villa Stein, in Garches, Le Corbusier's 1927 project, and Villa Foscari, Malcontenta, in Mira, by Andrea Palladio, 1550–60, where he demonstrates similarities for the fact that both are designed as single blocks, and by the relationships between their measurements and structural meshwork, where modular matching is clear. (Rowe, 1976: 5).

Thus, our investigation examines to what extent the classic Vitruvian compositional parameters are present in some examples of the history of housing in Recife, Brazil. For this, 4 representative examples of the history of housing in this city are analyzed, since these parameters become part of the architectural design procedures adopted in the country from the academies, in the 19th century.

These moments are the classicist, the art-deco, the modern and the contemporary, particularly with regard to the social spaces of these dwellings. Whenever possible, due to the ephemerality of the object, the composition of the interiors will also be analyzed, therefore, adopting as determinant the interior-exterior relation for the practice of interior architecture, it is observed that these were given as a result of the composition of the exterior, to a greater or lesser degree.

4 THE CLASSICISM IN THE HOUSE OF RECIFE

Until the nineteenth century, the constructions in Brazil from the arrival of the Portuguese in the sixteenth century were based on the classicism derived from the treatises that followed that of Vitruvius[8] in the Renaissance period, prior to the French writers[9] of the seventeenth century. With regard to the housing architecture, we investigate the degree of participation of classical guidelines that governed these projects until the introduction of the classicism

7. ordination, disposition, eurythmy, commensurability, decorum and distribution – free translation of authors.

8. Tratados de Leon Battista Alberti, Sebastiano Serlio, Giacomo Barozzi da Vignola, Andrea Palladio e Vicenzo Scamozzi. (Summerson, 2009:142).

9. Tratados de Philibert de L'orme, Roland Fréart, François Blondel, Claude Perrault, G.L. Cordemoy, M.A. Laugier, Jacques François Blondel e J.N.L. Durand. (Summerson, 2009:143).

of the imperial period, assuming that they were restricted to issues of tectonics, as well as to the constructive guidelines of the Municipal Postures. In the Badia House plan illustrated below, there is no link between its plan and the Vitruvian compositional principles adopted later (Fig. 1).

4.1 *Neoclassical housing*

The first dwelling analyzed is the current headquarters of the Academia Pernambucana de Letras, in the Graças neighborhood, Recife, a project of the French engineer Pierre Victor Boulitreau, built in 1860. Unlike the Badia House plan, it already demonstrates a compositional erudition of classic references. His plan is a square divided into 3 equal parts, examples of vitruvian ordenação and euritmia. It is also verified that the "golden rectangles" controlling the external composition are inscribed on the façade, as well as the concept of disposição, applied to its windows in arches distributed neatly, in composition with the arched portico that jumps from the central block, and which supports the upper terrace. It also has comensurabilidade, with its lateral wings, from the central wing as axis, worked in a balanced way. A decoro, referring to the identity bonding of the pieces between them "what is good" being taken into question, since there is nothing to break the unity of the environment, resulting in a harmony of it. The concept of distribuição does not apply, since its coatings, furniture and equipment came from European countries (Fig. 2).

4.2 *"Art-Déco" housing*

The second dwelling analyzed is the semi-detached residence in the neighborhood of Parnamirim, Recife. The "Art Deco" style, already in the first half of the 20th century, marks another moment in which a classical heritage is seen in its composition, with a pronounced front in analogy with a triangular pediment. Its plan is distributed between 2 floors, symmetrical, where one can identify the ordenação of the same from the insertion in 4 squares,

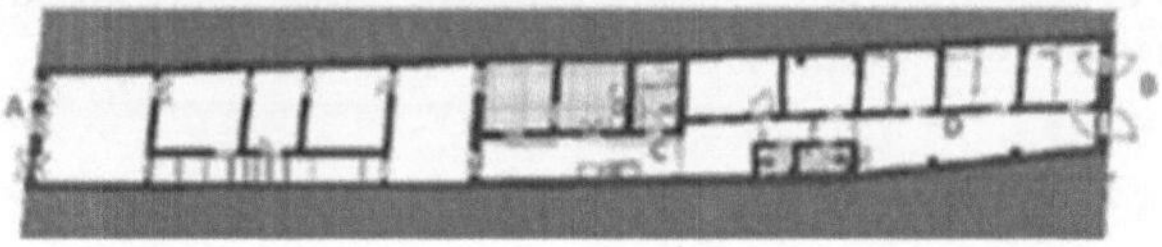

Figure 1. Ground floor of Badia house. (Source: Image courtesy of Prof. Ms. Pedro Henrique Valadares).

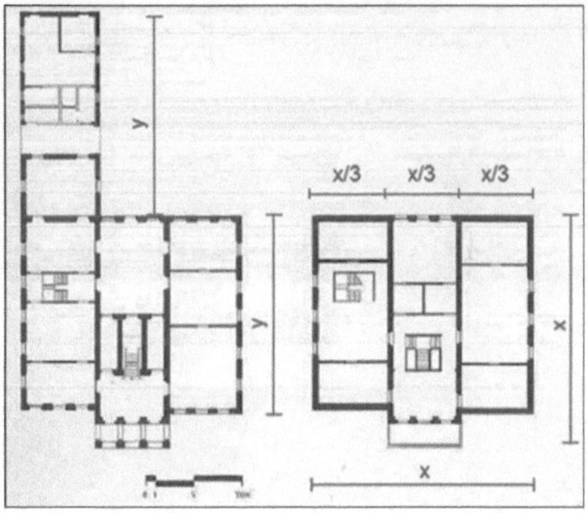

Figure 2. Ground and first floor of the residence of Barão Rodrigues Mendes, current Pernambuco Academy of Letters. (Source: Drawings of the authors on image found in Montezuma – org, 2008).

formed by a module that repeats in the salient part frontal and posterior to these squares. The disposição arrangement of the frames, leaking the bare cloths of the facade, typical of the style, produces an "elegant" and "pleasant" effect, translated by Vitruvius as euritmia. This option for a façade devoid of adornment, clean, was what was appropriate at the time, an example of decoro. From its symmetrical arrangement, from an axis where the ridge of the roof in two waters is located, one can observe the concept of comensurabilidade, or a correspondence between the parts that derive from this axis, balancing the façade. This residence does not identify elements that lead to define the concept of distribuição (Fig. 3).

4.3 *Modern housing*

The third residence is the Carlos Augusto Fernandes Residence, 1963, of the Portuguese architect Delfim Fernandes Amorim, in Casa Forte, Recife. The frontal façade, elaborated according to a proposal to enhance the appearance, has a gable that visually disguises the areas of the house to the street, an example of disposição, which refers to the proper placement of the elements and the resulting aesthetic effect. This gable, clad in stone, contrasts with the wood of the frames and the plaster of the masonry, exemplifying the decoro Vitruvian, using "what is good", which suits each case in particular. And because there is an understanding of the design practice of Delfim Amorim in this region, the concept of distribuição is identified, since it has always emphasized the local elements, both in terms of materials and in terms of constructive and cultural aspects.

The concept of Ordenação, as a proportion of the parts that make up the dwelling, both separately and in its totality, is connected to euritmia, that is, a proportionality between the parts that compose the work. These concepts are demonstrated in the plan presented in this house, where one can perceive the modular structure that generated the environments, all defined by modules, also verified in the façade. Thus, the plant presents an "elegant" exterior shape and a "pleasing" aspect, that is, the harmony of which Vitruvius speaks.

The concept of comensurabilidade or balance of parts of the work and the correspondence of a certain part with the others that make up the project, is identified here, due to the harmony achieved of the whole, and also in the layout released, supposedly by the architect, where from the furniture was arranged (Fig. 4).

4.4 *Contemporary housing*

Finally, the example of contemporary housing, VPRG 2005 project, in private beach condominium, in Muro Alto, Ipojuca. In this house the social environments, located on the ground

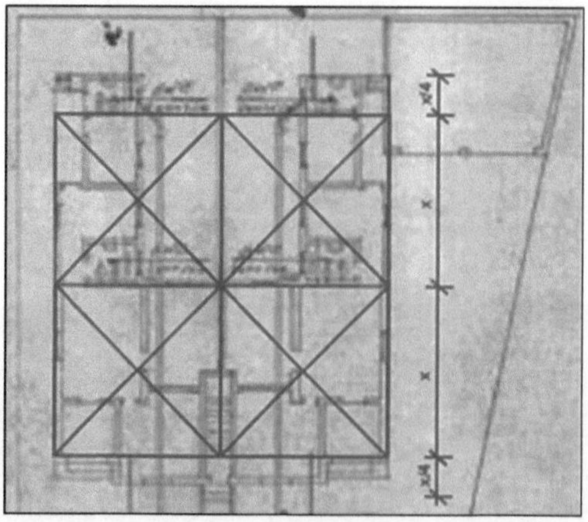

Figure 3. Ground floor of Art-Déco dwellings (Source: Drawings of the authors on the plan found in Barthel, 2015).

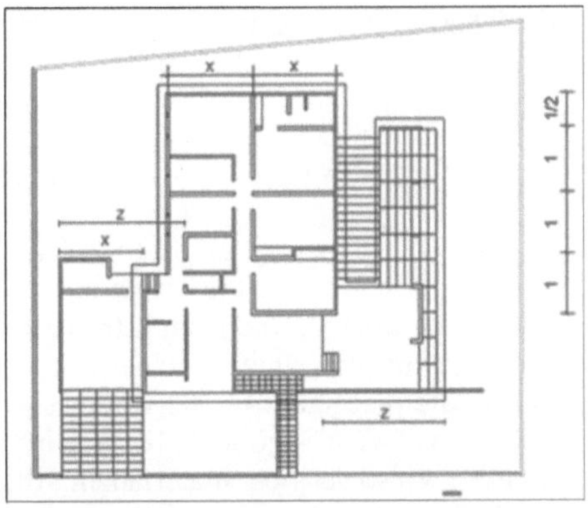

Figure 4. Ground floor Carlos Augusto Fernandes Residence. (Source: Authors' drawing based on plan of Pereira, 2006).

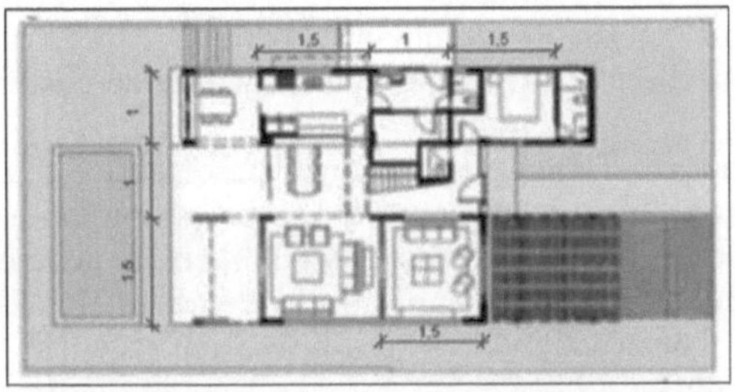

Figure 5. Ground floor Marcelo & Dulce Residence. (Source: Authors' drawing in a plan found in Muniz, 2012).

floor, are integrated with the kitchen cup and the outdoor leisure area, while other environments are more reserved. This aspect indicates the cultural permanence in contemporaneity, of the tripartition of the social—intimate—service sectors in the dwellings, although in this case we already see a relative union of the area considered as "service", the kitchen, with the social spaces. The Vitruvian concept of disposição is here identified, relating also to decoro, or to what is appropriate in each particular case, expressing itself according to local custom.

Thus, when we look at the plan of the house, and its modular layout, we notice the application of the concept of ordenação, a concept that merges with that of euritmia, which generates a proportionality between the parts of the work. The concept of comensurabilidade is applied in the geometric game of the facades, and in the arrangement in plan, generating in its interior also an arrangement with furniture arranged along an axis, and placed in such a way as to have a contemplative visibility of the external area, if this view as the background of a scenario. This background is also imagined if the observer is placed in the opposite direction to the outside, on the walls of the room, which become a background of classical, symmetrical, ordered composition for the disposition of all the elements that derive from it. And finally, the concept of distribuição is identified from the moment that architects make use of technologies and materials constructively dominated by the local community (Fig. 5).

5 FINAL CONSIDERATIONS

In the social spaces of the dwellings presented, although they were of different epochs and natures, and also of having been erected with totally different technologies and materials, each in its own time, were identified the Vitruvian classicist compositional principles researched in his Treaty of Architecture: ordenação, disposição, euritmia, comensurabilidade, decoro and distribuição.

The classical compositional principles extend from the facades and accesses to the interiors of these spaces, when it was possible to make the analysis, demonstrating the connection of a classical building to the composition of its interiors, which tend to have a composition also classic as a consequence.

The four analyzed dwellings had the geometry, as guiding element of their compositions, that is, the reference to the pure geometric solids, the regulating tracings derived from an initial axis, that governed the area to be worked, and from it the composition of the laterals, in view of a comensurabilidade, or equilibrium, which is achieved by ordenação of the elements, as well as a disposição, an adequate placement of the volumes, rooms and furniture, with respect to their interiors. Decoro, as "what is good," "the irreproachable aspect of works," "which suits each particular case," is also verified, materializing these projects "according to custom" and "in a natural way". The euritmia or harmony was achieved, starting from quite different proposals, but which, governed by the classicism of these compositional parameters, generated "elegant" and "pleasant" results, as Vitruvius names.

The concept of distribuição, among the four examples, was not reached in the first, of the neoclassical house, taking into account the historical practice of the local culture of valorization of what comes from outside, predominantly of the European and American countries, only reinforcing the question of the "image" intended to be passed by project observers, emphasizing the aesthetic dimension that is verified in the place.

It is also observed that all of them, still existing houses, although some very old, privilege the social spaces of the dwelling, the object of our investigation, both with regard to its privileged location in the lots, as regards its plasticity and its more generous design, with the use of coating materials and outstanding compositional elements. This leads us to conclude also that, due to the date of construction of the last house analyzed, the VPRG office, this habit is still a constant these days.

This small sampling of a much larger universe, with respect to the compositional principles of housing projects, has a character only as a sign of a hypothesis, which claims to be more substantial when the wider investigation is to be developed in the future.

REFERENCES

Barthel, S.G.A. 2015. Vestígios do Art Déco na cidade do Recife (1919–1961). Abordagem Arqueológica de um Estilo Arquitetônico. Tese de Doutorado em Arqueologia. Recife: CFCH, UFPE.

Janeiro, P.A. 2010. Origens e Destino da Imagem—Para uma fenomenologia da Arquitetura Imaginada. Lisboa: Chiado Editora.

Montezuma, R. – org 2008. Arquitetura Brasil 500 anos. Recife: IAB/UFPE.

Muniz, M.B. 2012. A experiência residencial na obra de Vera Pires e Roberto Ghione, 1998–2012. Dissertação de mestrado. Recife: Universidade Federal de Pernambuco, CAC.

Peixoto, G.R. 2000. Reflexos das Luzes na Terra do Sol: sobre a teoria da arquitetura no Brasil da Independência: 1808–1831. São Paulo: ProEditores.

Pereira, F.T.B. 2006. Residência Moderna Nordestina. Trabalho da Disciplina Habitação, Metrópole e Modos de Vida, ministrada pelo professor Marcelo Cláudio Tramontano. Programa de Pós-graduação em Arquitetura e Urbanismo. São Carlos: Escola de Engenharia de São Carlos, USP. Available in http://www.nomads.usp.br/disciplinas/SAP5846/mono_fulvio.pdf access 26.11.2017.

Rowe, C. 1976. The mathematics of the ideal villa and other essays. Massachusetts: The MIT Press. Available in https://monoskop.org/images/9/96/Rowe_Colin_1947_1976_The_Mathematics_of_the_Ideal_Villa.pdf access 10.02.2018.

Summerson, J. 2009. A linguagem clássica da Arquitetura. Tradução de Sylvia Fischer. São Paulo: Martins Fontes.

Tejo, C. et al. 2012. Uma História da Arte? Recife: Fundação Joaquim Nabuco—Editora Massangana.

Vitruvius, P. 2007. Tratado de Arquitetura/Vitrúvio. Tradução do latim, introdução e notas M. Justino Maciel. São Paulo: Martins Editora Livraria Ltda.

Drawing (...) City (...) Body, Dwelling on Earth – Janeiro (Ed.)
© 2019 Taylor & Francis Group, London, ISBN 978-0-367-02602-8

Walking: Drawing the city with the body

J. Ferreira Crespo
CIAUD, Lisbon School of Architecture, Universidade Lisboa, Lisbon, Portugal

ABSTRACT: This essay is a brief and hypothetical one. We seek out to draw parallels between the act of walking and that of drawing. Aiming to study the common ground between both these human actions and in what way these might be similar. We claim that both actions take place in contiguous territories which ultimately converge towards an idea of city. Such will be the chosen starting point of a path which we hope will lead us towards a better understanding of the relation between body and city.

Keywords: Walking, Drawing, Body, City

1 WALKING: DRAWING THE CITY WITH THE BODY

Our text will be a brief and hypothetical one. We wish to think about the act of walking as a way of inhabiting the world, as a way of drawing a path with our bodies. Aiming to explore territories, as one does when walking, or in Octavio Paz's definition:

"Following a path: reading a stretch of ground, deciphering a fragment of world." (Paz).

In our view, the first gesture of architecture is previous to any construction, rather architecture starts from an awareness of spaces and territories, done firstly by a recognition of what surrounds us and where we are. In this sense, walking might have been the first action to grant us that awareness.

"The path is made by walking." (Machado).

The act of moving our bodies throughout the land, seeking out ways to better understand it, might have been the first constructive action towards our inhabiting the world as humans. The first step of a long road which carried out a transformation of the world. There is, in our view, a transformative quality inherent to walking, which changes both the surroundings as well as the person walking.

"It's precisely the action on our own body that allows us to transform nature into landscape, and space into place. It all happens in an instant, in a hic et nunc where a new state of things is created, where things that were invisible until that moment, are now seen, and relations between things which weren't there before, appear. Space changes without yet having been physically transformed, and prior yet to the drawing of a line or the conception of an object." (Careri, 2017: 127).

In a certain sense, we argue that walking was, and still is, the first and ultimate act of drawing the city, carried out by our whole and entire body. It is not farfetched to think that the first human settlements, or for that matter the majority of our actual towns and cities, were initially structured in relation to paths that were made by walking. Still to this day, paths are of the utmost importance, as Kevin Lynch points out in his 'the image of the city':

"The paths, the network of habitual or potential lines of move-ment through the urban complex, are the most potent means by which the whole can be ordered." (Lynch).

If we were to compare the human body to that of the city, using anatomical images, we might imagine the flux of people walking as being the city's blood, and the streets the equivalent to vessels, veins and arteries which carry out movement. Henry Thoreau draws another anatomical comparison:

> *"It is the body of which roads are the arms and legs."* (Thoreau)

From a perceptional approach, walking is perhaps the most structuring action in the manner by which we relate ourselves to the city. The truest and most direct way of inhabiting it. In the words of Juhani Pallasmaa:

> *"I experience myself in the city, and the city exists through my embodied experience. The city and my body supplement and define each other. I dwell in the city and the city dwells in me." (Pallasmaa, 2008: 40).*

We project ourselves in and through the city, all the while, tracing out the city itself. Integrally—by means of our entire body and being. In this sense, walking is a kind of animated drawing (by animated we imply soulful. From the etymological sense of the word in Latin 'anima', meaning soul), ante litteram. The soul of the drawing is given by its movement, it is an ever-changing drawing. Our surroundings—and our perception of them—shift as we do, whilst walking. Even without drawing, in a strict sense by hand, we are doing so, in a broader sense, with our whole body. Not to differently to young Werther's own experience:

> *"I am so happy, my dear friend, so absorbed in the exquisite sense of mere tranquil existence [...] I should be incapable of drawing a single stroke at the present moment; and yet I feel that I never was a greater artist than now." (Goethe).*

There is a profoundly creative aspect to the act of walking through the city, in the sense that one finds himself participating in the continuous process of renewal and rebuilding of his surroundings, all the time, giving meaning to the city and thus to himself. In Francesco Careri's words:

> *"It's a body-to-body with the city. We cannot know space without first going through it with our body [...] The world becomes a permanent building site for the "re-cre-ation" of the body, a place of creation, action and transformation." (Careri, 2017: 124).*

One imagines the city in—and through—ever new ways, changing alongside with it. In other words, we change alongside the images we create of the city, for in the end, the image of the city is our own.

> *"The city returns nude images /from the profoundness of the being" (Lacerda, 1991: 85).*

We digress... However, in doing so we seem to be following the path the text is leading us to. A state of errancy (errancy is from the Latin word errantia: a wandering), possibly similar to that which lead to the construction of our cities, to the construction of ourselves. Curiously, to err, has a double meaning, its other, and most common, meaning is: to make a mistake, to be incorrect. If such is the case, then this would mean that we have mistook our path—but is this possible? for once one has taken a path it seems to be an impossibility to mis-take it—or rather, we are going astray. To this possibility we must remind ourselves that: errare humanum est...

> *"Space appears as a realm of objectivity, yet it exists in a social sense only for activity—for (and by virtue of) walking [...] In one sense, then, space proposes homologous paths to choose from, while in another sense it invests particular paths with special value. The same goes for angles and turns: what is to the left may also be sinister; what is to the right may also be 'right' in the sense of rectitude." (Lefebvre, 2008: 191–192).*

Thus, it is up to us to choose which line to draw and how to draw it. Such is the freedom of choice one has when walking through and in the city. The city which is continually rebuilt upon images that we actively draw from, into, and onto. In other words, by walking we draw our own city—mentally and physically, which is to say: Bodily. This may, and perhaps, should, be understood subjectively, however these considerations retain an objective aspect—which becomes clearer when one evokes the historical emergence of human settlements.

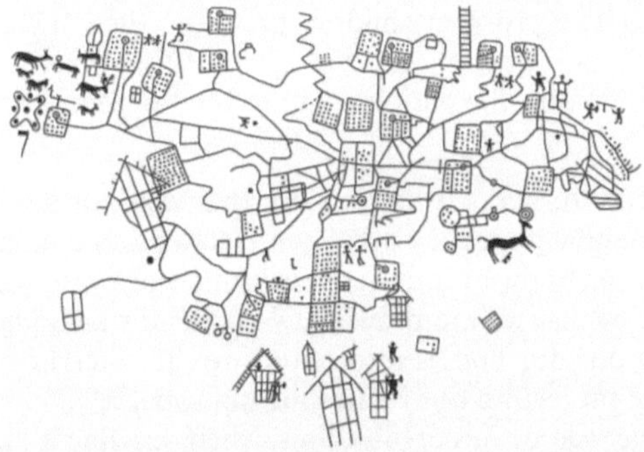

Figure 1. Petroglyph, Val Camonica, Italy, 1.200-700 BC. http://www.rcg.cat/images/50_10_01.jpg.

Figure 2. Richard Long, A line made by walking, England, 1967. http://www.richardlong.org/Sculptures/2011sculptures/linewalking.html.

> "The networks of paths and roads made up a space just as concrete as that of the body—of which they were in fact an extension." (Lefebvre, 2008: 193).

At the same time, each line we draw, or tread, unfolds a new territory. As Novalis would tell us:

> *"Every line is the axis of a universe" (Novalis, 1986: 17)*

Thus, and in this sense, might the act of drawing be one of discovery? Is the act of drawing, in a certain sense, continuous and contiguous, to that of walking? Might the root of drawing —or painting for that matter—reside in the discovery that our bodies leave traces? lines?

> "[…] we cannot imagine how a mind could paint. It is by lending his body to the world that the artist changes the world into paintings. To understand these transubstantiations we must go back to the working, actual body—not the body as a chunk of space or a bundle of functions but that body which is an intertwining of vision and movement." *(Merleau-Ponty, 2009: 19).*

By drawing a line are we somehow 'walking' a path?

Henri Michaux would come to write about the paintings of Paul Klee, stating the following:

"Lines walking around. [...] Travelers, lines that don't so much make objects as trajectories, paths. [...] The problem children have and then subsequently forget, the one they put into all their drawings at that age: locating things, leave here, go there, the distance, the directions, the path leading to the house as necessary as the house itself..." (Michaux)

Paul Klee, who we remind, said: "A line is a dot that went for a walk."

As for us, do we somehow go for a walk when tracing a line on a surface? If the line we draw is an extension of our body, or in other words, if by drawing we ex-tend ourselves, then yes, we somehow 'walk' whilst we are drawing. We extend our bodies in movement, we are continuously stretching out our bodies into/onto/through a territory—be it paper, stone or land. By drawing we are projecting our bodies through space.

What distance do we cover, or rather, un-cover, when drawing a line?

Pedro Janeiro wrote:

"The function of a pencil, when being used, is to draw a line. The cylinder of graphite which inhabits the darkness of my cedar pencil, in molecules, is a diamond (a carbon in its purest state); its function is to, through me, my body, find the light that flows within the contours of the things that build that-which-i-call-world. The graphite of my pencil, or the cubic centimetres of ink within my pen, may, potentially, build a line with millions of meters. From the point of my pencil or my pen, when pressed upon a surface, comes out, of the 'disordered world', to a 'world seen' by me; a world rebuilt through lines which define borders [...] Dots, centimetres, metres, kilometres of lines. A line that sleeps within my pencil, chaotic, without sense, inside a cylinder; but which my body, through it, by its own will, decides to awaken the things of the world." (Janeiro, 2012: 55 56).

Thus, by drawing—as by walking—we, potentially, un-cover a 'world'. Our world...

We might have been walking around in circles... Let us correct ourselves—a circle is too perfect to describe the line we have drawn out with this text—at best, our 'walk' has been labyrinthic. Nevertheless, we do believe that much may be found by treading this path, perhaps not at its end, but throughout its course.

"[...] in our nocturnal dreams, we unconsciously return to the life of our travelling ancestors. It has been said that related to man 'all is path'; if we are referring ourselves to the most remote of archetypes, it is fitting to add: in man all is lost path. To link systematically the feeling of being lost to all unconscious walking is to find the archetype of the labyrinth." (Bachelard, 1990: 163)

And yet we continue. Treading timeless—or rather, time-full—paths which are somehow bound to take us somewhere.

Figure 3. Petroglyph, Val Camonica, Italy, 75-500 BC. http://www.labolab.net/wp-content/gallery/laberintos-de-la-antiguedad/val-camonica.

"The city is the realisation of that ancient dream of humanity, the labyrinth." (Benjamin, 2002: 429)

And the path goes on, as we dwell on earth…
The whole night
as life
is brief
t' spell out the world
and your body
(Macedo)

REFERENCES

Bachelard, G. 1990. A Terra e os Devaneios do Repouso. São Paulo: Martins Fontes.

Benjamin, W. 2002. The Arcades Project. Cambridge, Massachusetts: Harvard University Press.

Careri, F. 2017. CAMINHAR E PARAR. São Paulo: Gustavo Gili.

De Lacerda, A. 1991. SONETOS. Venezia: Centro Internazionale della Grafica.

Goethe, J. W. 2009. The Sorrows of Young Werther. [online] Available at: http://www.gutenberg.org/files/2527/2527-h/2527-h.htm [18.2.2018].

Janeiro, P. 2012. IL TEMPO E IL DIVENIRE. (n.p.): Chiado Editore.

Lynch, Kevin. [online] Available at: http://www.miguelangelmartinez.net/IMG/pdf/1960_Kevin_Lynch_The_Image_of_The_City_book.pdf [18.2.2018].

Macedo, H. 2015. RESTA AINDA A FACE. Lisboa: abysmo.

Machado, A. [online] Available at: http://www.whitepine.org/noroad.pdf [18.2.2018].

Merleau-Ponty, M. 2009. O olho e o espírito. (n.p.): Vega.

Michaux, Henri. [online] Available at: http://referentiel.nouvelobs.com/archives_pdf/OBS1067_19850419/OBS1067_19850419_088.pdf [18.2.2018].

Novalis. 1986. FRAGMENTOS. Lisboa: Assírio e Alvim.

Pallasmaa, J. 2008. The Eyes of the Skin. Chichester, West Sussex: John Wiley & Sons.

Paz, Octavio. [online] Available at: https://en.wikiquote.org/wiki/Octavio_Paz [18.2.2018].

Thoreau, H. 2008. Walking. [online] Available at: http://www.gutenberg.org/files/1022/1022-h/1022-h.htm [18.2.2018].

Drawing (...) City (...) Body, Dwelling on Earth – Janeiro (Ed.)
© 2019 Taylor & Francis Group, London, ISBN 978-0-367-02602-8

The comic as an academic product: Presence and territoriality[1]

C.G. Abdalla
Departamento de Projeto, História e Teoria, Faculdade de Arquitetura e Urbanismo, Universidade Federal de Juiz de Fora, Juiz de Fora, Minas Gerais, Brazil

C. Santos
Doutorando em Arquitetura, Faculdade de Arquitetura e Urbanismo, Universidade Federal do Rio de Janeiro, Rio de Janeiro, Brazil

ABSTRACT: The aim of this paper is to analyse an exercise for a design discipline during 2015 and 2016 developed for undergraduates in Architecture and Urbanism at the Federal University of Juiz de Fora, MG, Brazil. Comics were developed by students in groups, and this activity was carried out as one part of a complete exercise for a design of a healthcare building for a Primary Healthcare Center. Certain arguments are used in the discussion, namely: (1) the difference between the natural and the cultural world; (2) territory (geographical and physical site), territoriality (social and cultural location) and landscape (individual perspective to organize ideas, concepts and design of thoughts); (3) perception; and (4) goods (in terms of: geometric materiality, physical materiality, interpretation of subjectivity and view of transcendence of the world). The health care policies (guidelines, rules, structures and so on) also are in the context of the paper. The discussion found that academic people express their points of view through those comics (drawings and speech balloons as well as the comic itself). And they do not belong to that location, meaning that they are not part of that territoriality. However, there are bodies—undergraduate students—in the place, but their culture and necessities are different in comparison to the local population itself.

Keywords: Comic, Urban Design, Healthcare Territoriality, Human Body, Presence

1 INTRODUCTION

The aim of this article is to write a concise essay about an academic exercise of comics. It was applied and developed with undergraduate students in the fourth period by the discipline of architectural design and urbanism IV, in the Architecture and Urbanism Course of the Federal University of Juiz de Fora (UFJF) during the years of 2015 and 2016. It should be noted that the importance of the article is in the discussion of the narrative of the comics, which made it possible to identify how the student, a person in academic formation, from one of the most significant areas of knowledge and production of the space of cities and their buildings, facilities and landscapes, constructs a discourse about the place based on his presence, body and consciousness, in an urban environment, naturally, considering the proposition of the academic contract and its student base of knowledge of the profession. In this process, local analysis is requested in urban themes of the city, such as transportation, the history of the neighborhood, education, health, etc. Students are put into groups of three, four or five and carry out the experiential research in loco and present it by means of drawings and texts in a sequenced story; naturally, each group is responsible for the development of one of the themes set for the class. The set of thematic comics, articulated and added to the whole contribution of the class, generates a single comic of the chosen territory, or post, for analysis.

1. The research ethics committee of UFJF—CAAE 57179616.0.0000.5147.

In particular, this exercise is carried out in a purely technical way, that is, in the territories allocated to the Basic Health Units in the city of Juiz de Fora, these territories being linked to the Brazilian health policy and in the context of monitoring of health infrastructure. Once the university students have obtained their initial knowledge, via a thorough broaching of the health system, its levels of care and, in particular, Basic Care, they are taken to an allocated territory to get a feel of the place where they will develop the design process, at another stage and in a conventional way; however, they will be limited, in the final stage, to producing an architectural draft design.

2 ARGUMENTS

This text is presented with the view that there is a notion of the world detached from the notion of nature:

"What we call "world" or "reality", says Husserl, is not a set of things and people, animals and plants and minerals existing within themselves and which our ideas represent by transforming them into objects of knowledge. The world or reality is a set of 'meanings' or 'senses' that are produced by consciousness or reason. Reason is "giving of sense" and it 'constitutes reality' not as the existence of beings, but as systems of meanings that depend on the structure of consciousness itself. Or, as Husserl explains, the reality constituted by transcendental consciousness or transcendental reason does not refer to the existence of beings, but to 'essences', that is, meanings. The essences are true, universal and necessary because they are constituted 'a priori' by reason itself. The meanings or essences are the content that reason itself offers to itself to donate meaning, for transcendental reason is a donor of meaning and meaning is the only existing reality for reason." (Chaui, 2014: 103)

From this separation of nature and the world we have a context that allows us to evaluate the reflection of the world of the student in the construction of the story, since it is understood that he is the subject and actor of the narrative in the comics produced for the exercise of the course. From this we will observe, in our reflection, space and time, both in the stories and in the drawings. Space, as a category, is perceived as in Saramarco (2008), that is, as an "instrument of housing". "Instrument" here is in the existential sense of being-in-the-world. A being that will be the centre of truth (Teixeira, 2006) and will allow, through the wording, the visualization of their presence as a subject in reading the place (the students and their groups in the construction of comics).

The separation, going further, allows a certain limitation of spatiality that refers us to different scales in the perspective of the reflections on the part of the territory given for analysis, such as:

1. The objective territory, which is mathematical, geometric and/or physical, abstract and yet measurable, universal and sensitive;
2. The place, understood as a subdivision of the space that "is organized, configured and allowed to be perceived from the things, from the objects that make up meaningful chains [for a being]" (Saramarco, 2008: 73); and
3. The landscape, unlike the two previous spatialities, is linked to the individual, and is thus seen by means of a judgment of the subject:

   "It is I who have the experience of the landscape, but I am conscious in this experience of assuming a de facto situation, of bringing together meaning through all the phenomena and of saying what they mean in themselves" (Merleau-Ponty, 2011: 355).

In relation to time, two dimensions, as Merleau-Ponty (2011) posits for the perception of a "field of presence", are analyzed in the comics:

1. The dimension here-there; and
2. The past-present-future dimension. It is said that time is perceived by the movement which "is a displacement or a change of position [in space]" (Merleau-Ponty, 2011: 360); however, the objective benchmark of transference is the subject (body and consciousness) and the understanding that the world is not only made of things (elements, objectives,

environments, regions and places, among many others). Thus we have many worlds and in these there are many times and many spaces.

The approach to space is complemented by the analysis carried out by Cassirer (2011), which deals with phenomena from the point of view of things in units of space:

> "... The existence of the thing is linked to the existence of these units of space. The fact that a thing is for us precisely this thing and the fact that it takes root as this thing is something that results, above all, from the fact that we designate its "place" in the whole of intuitive space. At every moment we assign it a definite place, and we return to the content of these places by gathering in an intuitive totality which brings to us the representation of the movement of the object as a continuous change, determined by rules. And just as, through this process, the thing binds itself to the position of other objects, it seems defined in "real" space, we also assign it a spatial "dimension" and a "form" as objective determinations" Cassirer (2011: 242).

In short, for a sensitive and sensual, perceived and cultural space, this article works with four objects:

1. Geometric–sensitive, can be determined by a precise spatial coordinate system in a pure space (x, y, z);
2. Abstract—sometimes called physical, sensitive and/or natural, has systemic, measurable, uniform characteristics that are observed by means of the pure space of geometry and the absolute space of the phenomenon;
3. Subjective—related to interpretations of the senses and sensations, which, as Coquet (2013) states, should not be confused with mere sentiment for the Kantian world, although every act of spatial perception contains in itself an act of mathematical deduction and mediation, given human nature, such as habits and customs. This object in space provides impressions that surpass this physical materiality that presents itself to the individual; and
4. Transcendental—it is not truly an object, since it is beyond the senses and sensations of the subject; its spatial and temporal materiality is not necessarily examinable with the same instruments as the others. By transcendence we can therefore understand that the subject of the analysis and his body of the senses and sensations may have different situations in terms of knowledge, either for substance in the objective rational world or in relation to others, in space and time.

In this respect the notion of subject is conceived for the reflection of the article. It is present in a certain place that is perceived, but is not homogeneous, knowing that "it is divided into determined zones and directions; however, each of them has not only a purely intuitive sense, but one of an expressive character of its own" Cassirer (2011: 255). The student, as a subject present in the territory of observation, sees himself and works there in two ways, with his physical body and also with a judgmental and creative conscience, which allows him to have a spatial perception and a mental vision, meaningful and rich in terms of myths.

Since we are dealing with the health system, we follow the principles and directives of Organic Law 8.080/90, which regulates Constitutional Articles 196 to 200 of CF [Federal Constitution]/88 (Brazil, 1990 and Brazil, 1988), and deals with the context of Brazilian health policy, which is based on a systemic, universal, timeless and intermittent vision. In a regional setting, this will be transformed, in the city, into territorial boundaries, called health territories. These territories can be observed, both physically in the urban environment (in the form or design of the city), and through their political-technical-scientific nature of health care, more specifically in Primary Health Care, when they are thought of as territorialities, this is, with aspects that take into account the life of the people, their cultural and social backgrounds, and the person in terms of human biology. In this aspect, the systemic vision of macro-national politics can be contrasted with the local daily action of the health teams and the perception of the population linked to the places where these territorialities of health are constructed. In this second context, which is local, there is an essential aspect in the approach requested of the students, that is, that they develop their own senses in loco regarding these territories and territorialities in health.

In Brazilian health policy, the city unit is an element of a hierarchical chain formed by political-geographic territories organized on at least three levels: national, state and municipal. In a state territory there are other equally hierarchical and regionalized arrangements, where the sanitary macro-regions, sanitary micro-regions and towns are found. In each town, it can be seen, depending on its size, that there is also a formal division into more local hierarchies such as the establishment of sanitary sub-regions. The municipal territorial subdivision within the Family Health Program is part of the national policy. In this formal structure, there is a systemic sanitary order based on the ecological epidemic view which is important as one of the categories to be addressed by the academic work of the comics.

This technical knowledge, which is considered as an abstract object since it is a category of theoretical approach and spatial boundary, is not, however, understood in the same way in the community, although the community experiences it daily and intuitively recognizes that there is a defining order of local actions on the health scene, which is established aiming at "scientific-political domination" and the establishment of a regulatory order for the territory by means of institutional organization, as could be done for other urban issues (Castells, 1983: 296), and which comes from an idea derived from the social control of health since the nineteenth century (Foucault, 1979: 95). Bringing such understandings to the context of the article, that is to say, they are elements that corroborate and exemplify the differences between the resident of the place, the one who belongs, and the student, who is present technically and in a discontinuous way, with a globalized viewpoint, considering that the subject of the place sees their needs in a particular way. In this, it is considered here that there is an experiencing of the clash, both by the senses in the points of view for the same object—the city—and in the aspect of time for the two groups of actors, since the measurements, observations, systematisations, analyses of the students are open, compared to the "closed" model of the local resident, given that for the latter there is a closed model (ascribed by the public system). Thus, it can be understood that analyses of qualities and quantities take place by means of different values between the subject who has the place as belonging, or who has the place as presence.

3 DISCUSSION AND RESULTS

We attempt to discuss, in the light of the arguments, how the student's formulation in his narrative is produced. It is first established that the assumption of the separation of world-nature is an inherent fact in context.

As a subject, the world view is different from the objective logical reality pre-existing in the place, even though it is located at the centre of the existing scenario. All the comics construct their own scenarios of development of the story, where there are no geographical limitations of territories in relation to that established by what is allocated in the organization of the health space (see Figure 1).

Figure 1. The territory that is extrapolated in relation to the allocation to the basic healthcare unit (authors documentation – 2015 undergraduate students working with the subject of social aspect in the locality).

The university student, therefore, is master of himself, which consequently gives him a degree of freedom in his narrative in the comic. According to this view, following Coquet's thinking (2013: 102–109), where the presence of these bodies in space allows a confrontation of the logical space (in this case, Euclidean, physical and given) with the discursive space, of the clash between the I and the other (the one unknown to me and to whom I am unknown), of the phenomena, of the person's presence in the happenings to be presented as part of reality. In this, we cannot think that it is possible that the narrative of the stories is developed by a logic of the objective reality; it can even be said that it is an act of purely intellectual significance produced in great part by the experience of the world brought by the student in contact with his experimentation in the territory of observation, the scene of action (see Figure 2).

From a perspective of organization and analysis of the reading of the comics, here seen in their formal entirety as a discourse carried out by the students, even though in groups, it is understood that there is in them a syntagmatic order that allows us to recognize paradigmatic principles in a narrative provided by part of the formulations drawn and textualised in the comics.

In the territory at least two dimensions can be observed: time and space. The places of these comics are here considered influential, here established as scenarios of events, where, then, the city itself and its various scattered elements, which extrapolate the particular place, are naturally posited as data of analysis; more than this, they can be understood as catalysing elements for the determination of parameters of the discourse in the students' work (comics). We emphasize that the city itself is incorporated as a significant element, that is, which determines the situations, even if materially and physically fettered to the place, but goes beyond the physical territory of observation. In this case, it is apprehended or brought to the narrative by means of the use of objects such as: formal architectural and urban elements, equally formal aspects of its geography, historical and cultural data, among others (see Figure 3). There is in this a subject in the formulation of comic books which attributes a grammatical context that is convenient for it and does not necessarily reflect what is given by the sense of the body in the pure space of its reference, in other words, of the sensitive and sensory presence and experimentation of the environment in the territory.

Another important element to be noted in the discourse of the comics projects discussed here is time. Just like what has already been said [about place], it is also filled with objects that are not consistent with people's physical reality at the time of observation. In this sense, scholarly times (brought, for example, from knowledge of architectural styles), historical times (also introduced by other knowledge), imaginary times and also culturally given times can be gathered from what was proposed. However, it is with the presence of these subjects that it is also possible to say of the body in the territory, in the here and now, that they are

Figure 2. An important historical man for the development of the city and for the neighborhood called Mariano Procópio is a ghost and an actor in that comic (authors documentation – 2015 undergraduate students working with the theme of culture in the locality).

Figure 3. A creature that is a capybara (an animal that has been protected by law and that it is common to be found throughout the city of Juiz de Fora), it turns into a character with human characteristics in the narrative of this comic. Its function is to present the locality. It begins with its own natural habitat and reaches the important patrimonies of the city that means not only in the neighborhood. All this occurs as part of a walk with your child (authors documentation – 2015 undergraduate students working with the subject of heritage in the locality).

Figure 4. Narratives in student viewpoints: they are talking many times about your own lifestyles. For example, you can see in the frames above that the grandmother is in a living room watching the television set and in the next frame that there is one student running to the school van (authors documentation – 2016 undergraduate students working with the subject of education in the locality).

capacitated for the events that they narrate, as individuals or groups in the comics. This encounter allows for a two-order clash; between their personalities (construction as individuals) and what is found there; a clash of worlds. A clash between the differences in worldviews of what is pre-existing in social, cultural, historical aspects, among others, in a given place (in the sense of objective reality), where what the individual that I am sees from there and how it is that "Others" see the same place, that is, how people interpret their own formations of individual values as well as the judgments they make about things (see Figure 4).

The comics analyse and narrate the problem of place in the universe and from the perspective of the student, even when they want to see things from the point of view of some actor proposed by the narrator to speak (teacher, mythical animal, etc.). This can be observed by the quantitative measures that are presented, that is, by studies of real categories based on systems that are insufficient to grasp the formal structure of the place and also by the time of both the experimentation and the academic process to carry out in-depth measurements that would make it possible to understand phenomena of interest to the architecture to be designed. Students work on health systems independently, without explicit focus on their logic of politics or technocracy/political and technocratic logic and having as their vanishing point (substance of drawings and texts) the elaboration of architecture and comprehension of what they consider fitting in their views of urban problems. For example, the difficulty of crossing an avenue to reach the site of the Basic Health Unit, or the volume and quantity of public transport in the neighborhood. This understanding of the city is still more fragmented

Figure 5. Narratives in comics: A dialog between the architectural design professor and the engineer Mariano Procópio in 1915 talking about the urban design of the neighbourhood (authors documentation – 2016 undergraduate students working with the urban design in the locality).

by the academic contract that categorizes things into specific themes to be developed in isolation, in the first instance, by each of the groups. With this in mind, we can see that there is a rather shallow analysis of the problem of the sense of belonging, given the shortage of discussion and parsimony of concepts and scientific knowledge that use the city object and the health structure itself for the site (such as an epidemiological and social view of the place), although there is a strategy of the narrator to act because of circumstances and rules that he does not control.

Faced with this, three strands are open to the narrative:

1. One of bypassing for the narrator, who establishes a relationship between the whole and the parts among these subjects; this allows students to act with their own individualities and potentialities; more than that, even without communication during part of the process, that is, by means of "subcontracts" agreed internally by the members of the groups;
2. Another of quantity, contrasting the collection of objects with the correlation of unity to the comic, which allows them to approach elements in a complementary way, as categories such as: inside and outside (spaces and times) of the territory and of the object itself of the academic exercise, of characters of actions, of the use of metalanguage, etc., therefore related to non-personifiable situations; and
3. A logical model of action consisting of two planes of integration:
   a. Universal and integral, or mathematically dealing with entire amounts (understood when elements that create connections—be they direct, partial or indirect—of identity between the territory and the city, such as, for example: historical characters, history itself, or elements of the nature of the site in correlation with elements of the city); and
   b. Discrete (collections of objects in classes and of a finite form, for example, number of inhabitants, or quantity of services in the territory and outside it), or of paradigms and semantics (in the context of the narrative) for the stories (correlated drawings and texts), the page layout structures, etc.

It can be seen that the subject(s) act(s) with the purpose of completing the processes, sometimes prospectively, sometimes retrospectively, sometimes looking towards the future to be carried out in the activity of the discipline, sometimes with the achievement of the past in the attempt to position themselves or understand the present. There is a sequence that is established, then, based on the academic contract, which is qualified with the action in the field and the technical-scientific and theoretical support and which is finalised sanctioning what is significant, in the view of the person who recounts and structures the logic of the narrative about the territory.

Taken together, by a reflection when looking at the production of comics of the class as a whole, we cannot, from all that has been said, expect homogeneity of thoughts, which allows the construction of the story sequenced by the parties in the formulation of the class comic. Thus, these objects found there in the field of language are words to be interpreted by individuals, simultaneously, mythologically, subjectively, sentimentally and sensitively. They objectify, subjectivate and transcend by means of presences in the territory to develop the comics. Moreover, taken together, as can be interpreted by the totality of the class comic, there arises a certain universal awareness of the spontaneity of the spirit that reveals the senses and knowledge particular to the nature of that politically and geographically arranged space. Thus, we can see that a trajectory opens for the student to give continuity to his architecture, conciliating a language of conception of the building, technically and stylistically speaking, tied up with the arrangement of the place already established. Here the words of Cassirer are appropriate:

> "Language develops as a sphere peculiar to the human being, derived from a universal drive tending to sensory and mimic expression, which is inherent to it, not as a result of reflection, but rather unconscious and involuntary" Cassirer (2001: 128)

In the form of the instructions proposed by the discipline of architectural design and urbanism, the exercise, seen as a contract, is imposed as an object that leads to a way of acting that is particular to the scholastic view, which entails the inclusion of another object of subjectivities for the student in his way of looking at the scene of action, whether in the observation, the analysis or the construction of narratives. In this respect, the experience undergone by the student-subject is also seen to clash with certainties and uncertainties, both from his knowledge of what is to come in, architecturally speaking, the academic exercise, and from the views of facts that they see with the apprehension of the place of the intervention for the development of a design of the object built (mainly parts of the architectural object and built as elements of the city and which will be the next step in the logic of the academic discipline). The infraction, in this conflict between what is observed in loco, what is thought of and what is requested by the academic body, seems to us to have led the subject to the need to establish a sovereignty over his observation and history, which takes place in terms of mastery of his body in space, translated by where he chooses to make the trajectory in the geometric space, by what his look and other senses search for in terms of objective and subjective elements that can add value to the future, including perhaps something placed at the scene of the place by means of objects that are not data which are present there. This future contract is also seen as a discontinuous element in this process, since it is systemic, that is, independent of the daily life of the place. The student, by taking an active role through the development of his own story, becomes the protagonist in the exercise of the comics and then in the process of the semester as a whole. This happens when he overcomes the barrier of what is functional and is able to develop an urban reflection about the place, as well as understanding that responding exclusively to the academic contract can transform him into a non-subject, as he will not overcome challenges that are inherent to the process. As a non-subject he will only be a producer of architectural responses to the issues imposed by the norms. In order to stop being an arranger of architectural layout and urbanism and become an actor committed to a new scenario (through his own architectural object), the student realizes that it is important to be aware of and recognize the other subject's place in the installation of his architecture.

> "The experience of architecture as an active (everyday) practice and unstable production of the social-space of relations... an act of presentiment of diversity and territorial differences... a practice of invention of a possible territory..." (Correa, 2017: 122)

Then it makes the agent who is a student and producer of the architectural space of a Basic Health Unit need to know the relevance of his work to collectivity and to citizenship in the allocated territories. To achieve this, in the language used by the student or group, he or she acts with a formal (non-subject) I, sometimes with a substantial I (subject). His body in the space of the scene sometimes is flesh (matter), sometimes is consciousness (concept and judgment), sometimes is unconscious (it is in another scenario, far from the territory and place). The I (individual, person, character of the narrator) acts as a link in the present, but constantly with links to the past and to the future, and always limited to what is proposed and delineated by the territory. There is a reality in the facts, but the reality of the presence and not necessarily of the belonging, acts bearing in mind that their interventions are limited, and their bodies, directly, do not belong culturally and socially to the observed territory. In this sense, the article concludes that the comic is positive in order to instigate the reflection of the student about the role since the city fears the limits with which he can work his body in the identification of territories and territorialities.

REFERENCES

Cassirer, E. 2011. A Filosofia das Formas Simbólicas: Terceira Parte: Fenomenologia do Conhecimento. São Paulo: Editora WMF Martins Fontes.

Cassirer, E. 2001. A Filosofia das Formas Simbólicas: Primeira Parte: A Linguagem. São Paulo: Editora WMF Martins Fontes.

Castells, M. 1983. A Questão Urbana. Rio de Janeiro: Editora Paz e Terra.

Chaui, M. 2014. Convite à Filosofia. São Paulo: Editora Ática.

Constituição Federal. 1988. Brasil [online] viewed: 28/1/2018 Available at: http://www.planalto.gov.br/ccivil_03/constituicao/constituicao.htm.

Coquet, Je-C.M. 2013. A Busca do Sentido: A Linguagem em Questão. São Paulo: Editora WMF Martins Fontes.

Corrêa, A.M. 2017. O Segredo do Arquiteto: Perdão por Não Lhe Abrigar. Belo Horizonte: Editora UFMG.

Foucault, M. 1979. Microfísica do Poder. Rio de Janeiro: Edições Graal.

Lei 8,080. 1990. Brasil [online] viewed: 28/1/2018 Available at: http://www2.camara.leg.br/legin/fed/lei/1990/lei-8080-19-setembro-1990-365093-normaatualizada-pl.pdf].

Merleau-Ponty, M. 2011. Fenomenologia da Percepção. 4ª.Ed. São Paulo: Editora WMF Martins Fontes.

Saramarco, L. 2008. A "Topologia do Ser": Lugar, Espaço e Linguagem no Pensamento de Martin Heidegger. Rio de Janeiro: Ed. PUC-Rio/ São Paulo: Loyola.

Teixeira, S.M.P. 2006. "A Noção de Habitar na Ontologia de Heidegger: Mundanidade e Quadratura". Dissertação de mestrado, Universidade Federal da Bahia; Salvador [online] viewed 5/8/2016; Available at: http://www.gephermeneutica.ppgf.ufba.br/trabalhos_concluidos/SoniaMariaPlatonTeixeiraAnooodehabitarnaontologiadeHeidegger.pdf).

Drawing (...) City (...) Body, Dwelling on Earth – Janeiro (Ed.)
© 2019 Taylor & Francis Group, London, ISBN 978-0-367-02602-8

Urban vitality in Madureira: Field observation and recording

J. Rocha & M. Costa
Postgraduate Program in Architecture and Urbanism, School of Architecture and Urbanism, Fluminense Federal University, Niterói, Brazil

ABSTRACT: This essay proposes a model of field observation and recording guided by the characteristics and conditions of the use of the place, which are identified according to the urban vitality of the place. In this work, we have assessed the traditional Madureira neighborhood, located in the suburban area of the city of Rio de Janeiro. This neighborhood was part of the process of constructing a "global city" in Rio de Janeiro, based on the re-qualification of public spaces and the strengthening of local culture with the objective of preserving values, communication among users, and their self-identity. The main hypothesis is that, in spite of the many efforts made by the public power to build a "scenario-city", idealized within the search for the so-called globalization, the process has, in reality, resulted in the construction of a different city, which is the result of the manifestation of its population. This can be recognized in Madureira by observing how the reminiscent spaces of road constructions were appropriated, and how they have become part of this identity, awarding this space a very peculiar vitality. The methodology consisted of a theoretical-practical approach combined with the theories proposed by three important theoretical references in the field. This approach allowed the systematization of three basic characteristics of observation and recording on the selected object. The initial hypothesis was verified by a "city reading" carried out in the neighborhood of Madureira.

Keywords: Urban vitality, Identity, Global city, Madureira, Rio de Janeiro

1 INTRODUCTION

Cities are the combination of transformations produced by society over a physical space. More than ever, the contemporary city is characterized by an increasing offer of multiple spaces of diverse complexities. These spaces can vary according to their purposes, aesthetics, the ability of their inhabitants to perceive this space internally and externally, as well as their relationship of belonging and identification. Cities are existing spaces that, in a given moment, are appropriated ways that differ from their original purpose. The objective of this study is to propose a "city reading" through the monitoring its daily life. The assessment was guided by observations and recordings to identify its urban vitality. Here we propose to read one of the multiple cities that exists within the city of Rio de Janeiro and that is not part of the showcase display of the city, built within the contemporary context of production of "world-cities" or "global-cities".

The study was carried out in the neighborhood of Madureira, located in the traditional suburb of the city of Rio de Janeiro. This region was part of the process of creating a global city, but not in the sense of the requalification of its public spaces and the strengthening of the local culture to preserve their identity. The area went through great transformations with the implementation of the express corridor for the Transcarioca BRT (Bus Rapid Transport) and suffered great impacts from the road construction. These transformations left the region with several residual spaces under the new viaduct and its access ramps.

There is a significant volume of positive criticism towards the idea of producing new models of cities based on the argument of the development of globalized economies. These have been presented through various marketing strategies as a viable alternative of great urban projects that would transform these cities into world symbols, in contexts that are often attached to a major world sporting events. There are also a number of negative criticism discussing how damaging it can be to reproduce this urban model, which has been applied in several parts of the world, especially in regard to local identity. The projects lack the knowledge about the original lifestyle of its inhabitants and are commonly distanced from their genuine characteristics in the praxis of building and rebuilding spaces. In the trail of policies to build "global cities", such as Rio de Janeiro, the cities have been led to forge images or scenarios based on exogenous, non-local models to the detriment of preserving identities which were woven by several generations.

In this sense, what is being been discussed is the production of model-spaces which configure contemporary scenographic cities. These actions have been oriented by projects of urban interventions as a process of creating "places of success", reproducing international models of urban requalification to reorient the city toward its insertion into a global and competitive economy. In this context, the areas chosen as locations for urban interventions, and that respond to this ideology, are those that have greater visibility (at least as a product of urban marketing), tourist potential, and that are attractive for international investments, encouraging the entry of foreign capital. For a long time, other questions were left out of the debate: what about the rest of the city? Why is there no debate on the construction of a city that, although not scenographic, could be much larger than this remodeled portion?

In view of this persisting reality, there is a disagreement in relation to these debates, especially considering its focus on this contemporaneous model of production of cities. This restlessness has led us to take a closer look at this city. In this essay, we advocate that, despite the many efforts towards building an idealized "scenario city", in search of a so-called globalization, it possible to identify the construction of a different city within, which emerges as a manifestation of its local population, as can be observed in Madureira.

To support this hypothesis, we have adopted a theoretical-empirical approach, in the sense that, it was necessary to carry out a literature research to discuss the questions which emerged from practical/empirical problems. These empirical questions were raised from the experience and the findings obtained by the first author as an in loci researcher and observer. Parting from an empirical reading of the city and supported by the works and ideas of Jacobs (2000), Lynch (2007) and Gehl (2013), the objective was to develop our own theoretical reflections, organized according to the observation and recording of the urban vitality of Madureira. This approach allowed us to produce an updated overview of the city downtown region and also of the Madureira neighborhood, associating identity, centrality and activities established by its community.

2 THEORETICAL REFERENCES

The main theoretical references of this work were Jacobs (2000), Lynch (2007) and Gehl (2013), which guided the systematization of certain categories of space observation and reading. All these authors reflect upon the same object—urban vitality—from their own perspectives and forms of perceiving it in space.

When analyzing the Death and Life of Great Cities, Jane Jacobs (2000) demonstrates that urban vitality is related to the diversity of uses and the vibrancy of places in consonance with social interaction. According to Jacobs, the promotion of the diversity of uses of the city is the best and most effective way to fight the monotony produced by the rigid, sectorized and mono-functional modernist urban planning. This diversity would also promote security, attractiveness and interaction among city inhabitants. Jacobs observes a serious crisis in urban social life, whose impacts have reverberated in the urban spaces of the great metropolises: physical barriers, urban voids and insecurity. The immediate consequence of this urban picture is the death of spaces of sociability, or places of urban vitality, which are characterized by the natural promotion of human encounters.

Kevin Lynch (2007) discussed the "Good City Form" from the perspective of the relationship between a "good habitat" and the "biological well-functioning" of the individual. The author presents a set of "performance dimensions" for the spatial form of the city: vitality, sense, fit, access, and control, all of which refer to a set of qualities. Among the five basic dimensions proposed by the author, vitality is presented as "the degree to which the form of the settlement supports the vital functions, the biological requirements and capabilities of human beings—above all, how it protects the survival of the species" (Ibid., P. 117). This dimension, therefore, is related to the definition of a suitable terrain for life, with fundamental qualities for health, good biological functioning and survival: (i) sustenance (adequate supply of food, energy, water, and air, and a proper disposal of wasters, adequate to sustain life); (ii) safety (a physically secure environment where hazards, poisons and diseases are absent or controlled); and (iii) consonance (between the spatial environment and the "basic human biological structure of the human being") (p. 119). Thus, in his work, Lynch discusses how environmental characteristics are connected to the health and the biological functioning of human beings.

In his "Cities for People", Jan Gehl (2013) resumes the discussion proposed by Jane Jacobs in 1961 on the modernist urbanist ideology, particularly with regard to the rigid zoning that separates the uses of the city, and that is associated to the increased car traffic. The author reminds us of the consequences of this process, and that were pointed out by Jacobs: lifeless cities devoid of people, that is, the end of the urban space and the life of the city. Fortunately, Gehl points out that in recent decades (this reference dates back to its first edition from 2010) there has been an effort to create better conditions for pedestrians and for urban life at the expense of the car traffic. On the other hand, he affirms that, unfortunately, this is not a latent process in all parts of the world and is more visible in the most economically developed areas. In emerging countries for example, due to an accelerated increase in car traffic, there is a dispute being waged over the space, producing less suitable conditions for urban life and pedestrians (Ibid, 5–6). The key to the future would lie precisely in addressing the issue: prioritizing the need of citizens, which is translated—into what Gehl denominates a "human dimension" that can be reached through four "key objectives": lively cities, safety, sustainability and health (2013, p. 6).

In this essay, we have focused on the key objective "lively city", which, according to Gehl should invite people to "walk, bike and stay in the city space", developing social and cultural activities. In other words, "the lively city also needs varied and complex city life; where recreational and social activities are mixed with room for necessary pedestrian traffic" (Ibid, p. 63).

According to Gehl, there is a close relationship between the possibility of a livelier city and improved urban spaces. In this context, the author argues that in order to have better urban spaces which would be able to promote more liveliness to the cities, it is necessary to consider the connection between specific aspects, such as the degree of concern with the human dimension and the usage and quality of the public space. Considering the quality of the public space, Gehl systematized specific criteria[1] to analyze this aspect, which can be used as a guide to analyze the city.

In view of the theories and ideas of the three authors previously mentioned we have proposed categories that may assist the reading of the city, specifically its day-to-day life, understood as vitality. There is no doubt that since the pioneering publication by Jane Jacobs in 1961, the term urban vitality has been the subject of discussion by scholars, urbanists, and urban planners. As a result of these debates, several features are considered as criteria or parameters of analysis of how and how much variation can be exerted over urban vitality.

---

1. [The criteria for urban space quality defined by Gehl (2013) are: I) protection: against vehicular traffic and accidents, against crime and violence and against unpleasant sensory experiences; II) comfort: invitations for walking, for standing and staying, invitations for seeing, for talking, for play and recreation, for exercise; and, III) delight: scale, opportunity enjoy the positive aspects of climate, positive sensory experiences].

Based on these authors, we have listed three categories that are characterized by their multiplicity or assortment of the presented variables:

i. **Environmental and spatial diversity:** refers to the variety of public spaces, qualified or not, which fulfill their primitive function (the purpose for which it was created), or that is being reappropriated in some other way. A second fundamental aspect is the identification of users or groups of users for the characterization of social aspects;
ii. **Diversity of accesses:** refers to the various possibilities of communication and fluxes through modals of circulation. It is necessary to characterize the network created from these identified connections between various points of the city;
iii. **Diversity of activities and use:** refers to the variety of economies, such as commerce and services, formal or informal.

In this work, we have used the neighborhood of Madureira as a spacial unit to make the above categories manifest, which will be further analyzed.

3 CHARACTERIZATION OF THE MADUREIRA NEIGHBORHOOD: HISTORICAL, ECONOMIC AND CULTURAL ASPECTS

In the last decades, cities have been attempting to insert themselves into a globalized world, transforming their spaces, most often as within a project of construction as a host-city for major world events. In this context, cities are planned to forge scenarios or images, based on exogenous, non-local models, to the detriment of an existing city that is built on a daily basis by their inhabitants. A city that does not serve this purpose, since there is no architecture signed by some "stellar" professional, does not offer possibilities for the tourism model that is in vogue at the moment, nor it becomes the object of analysis when discussing the production of global cities.

Madureira is a neighborhood located in the North Zone of the city of Rio de Janeiro. Its development in the late nineteenth-century (1890) was connected to the implementation of the railroad and the inauguration of its train station, which allowed the city to expand towards its suburban areas. The occupation of these areas was linear, following the railroad track and had greater concentration around the railway stations. Following this initial moment, new streets were opened perpendicular to the railroad, consolidating the neighborhood centers (Abreu, 2006).

In the 1940s, the nuclear city found itself in an intense process of expansion, witnessing the emergence of new centralities and intra-urban restructuring of the original spatial organization of economic activities based on the "center-periphery" model. According to Corrêa (1997), this restructuring resulted in a process of decentralization and formation of commercial sub-centers, identified as areas with great concentration of trade and circulation of goods and money, such as Madureira.

According to Bernardes (1990) until the 1950s and 1960s, Madureira was still characterized as a simple residential suburban area. Since then it has had an extraordinary development and it is nowadays positioned as a "sub-center", mainly due to the following factors: (i) its geographical location bordering a densely populated zone and also a region going through full expansion, which does not have the necessary services for an increasing population; and (ii) its geomorphological condition, since it is located in a bottleneck area, between 2 hills, where there are two trunk railways (Fig. 1), and from where a large number of connecting roads leading to suburban districts part, such as Jacarepaguá, in the West Zone of the city. Besides these conditions, the area has an "abundant and varied trade, where a large number of commercial branches from the city center [...] are represented, attesting to the vitality of Madureira as a commercial sub-center" (Ibid, p. 131). Bernardes (1990) also states the transformation in the physiognomy of the streets close to the train stations, which have been transformed from residential to commercial areas. In conclusion, the author asserts that the neighborhood can be considered "a very important regional sub-center, owing its vitality to

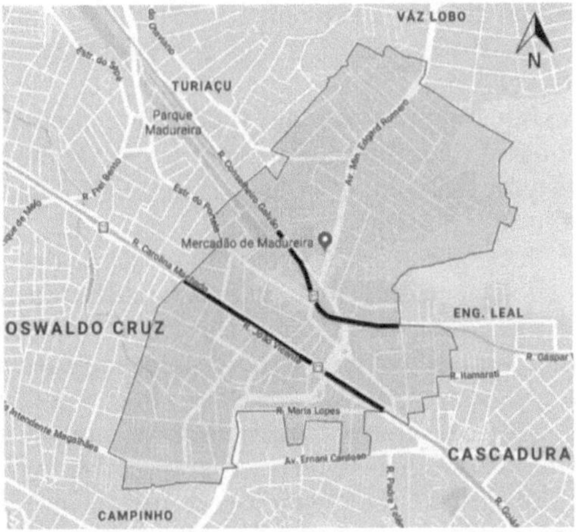

Figure 1. Map of the neighborhood of Madureira. Source: Adapted from Google Earth. Accessed on: 28 Jan 2018.

the primordial fact of its position of contact between the urban zone and the suburban zone" (Ibid, p. 132).

Since then, and in the following decades, Madureira has presented itself as an extremely relevant sub-center in the context of the urban dynamics in Rio de Janeiro. It is mainly associated with the growth and dynamism of commercial activities, especially in the retail market of clothing, personal and domestic articles, and the provision of services such as medical specialties, dentists, lawyers and accounting offices, among others. It can also be observed that this "third sector" is composed by an informal commerce that has become characteristic of this area: hundreds of "camelôs" gathered around the vicinity of the Magno/Mercadão de Madureira Railroad Station, or scattered on the sidewalks, walkways, ramps, stairs, bus stops and train stations. All of which produce in the neighborhood an identity connected to a place of popular market.

Madureira is also a place of rich manifestations of popular culture, such as samba: it is where the rehearsals of the Samba Schools of Portela and Império Serrano surpasses the limit of its blocks, and spreads to the adjacent streets; or the jongo da Serrinha, which is not only a representation of an African originated dance, but also promotes work and income opportunities through the preservation of culture and memory; or the Baile Charme in the Madureira Viaduct (Viaduct Negrão de Lima) that takes place every Saturday night in the Cultural Space Rio Hip Hop Charme, a strong representation of the black identity.

Madureira is a space of symbolic representations and rich in meanings. It expresses social structures, translated by the spatiality, materialized by identity that goes against the current official discourse produced for the city, which increasingly produces a homogenization of global landscapes. We believe that it is through the local identities that it is possible to produce resistance in face of the primacy of the globalized urban scenario. Therefore, we advocate that, despite the great efforts to produce a globalized city-scenario, underlying it, is the construction of a diverse city presented as a manifestation of a local population with the discussed urban vitality.

4 SELECTED CATEGORIES AND EMPIRICAL APPLICATION

Based on the presented characteristics we have carried out a "city-reading" of the urban spaces in the central area of the Madureira neighborhood. In this work, the chosen approach does not measure certain variables, rather, it seeks to identify phenomena and dynamics in these spaces, in order to determine greater or lesser urban vitality. In this sense, the following diversities and peculiarities were observed:

i. **Environmental and spatial diversity:** The center of Madureira is an area where there are almost no qualified public spaces, that is, designed to meet the expectations of its resident or floating population, allowing them to stay in the place, safely and comfortably. The Praça Magno (Magno Square) is an example. It offers no possibility of rest or encounter. There is also no adequate urban forestation, since trees that provide shade and a better thermal sensation are almost absent, even on sidewalks. Qualified spaces were identified only in the skirts of this urban center:

a. Praça do Patriarca—located beside the Campinho, an area that was part of the requalification in the second stage of the Rio-City Program developed in the 2000s, and that is designed to favor social practices. It has a playground (much used by children in the area), a society soccer field (where young people queue to participate in a match), exercising equipment (much sought after by the elderly in the morning and late afternoon), a forested area (with big trees that minimize the high temperatures) and the Clube de Malha, which existed before the intervention and received an appropriate field (due to the tradition of this sport in the region, the court was included in the scope of this work). It can be observed that these social practices that have qualified spaces post-intervention, favor sociability bonds and confer meaning to the place;

b. Madureira Park—inaugurated in 2012, it occupied an area where a transmission line of the energy company Light passed through, and was later occupied by a precarious community of approximately 900 families (Favela Vila das Torres) that lived below the line. The park has a linear configuration that is showing a perpendicular development in relation to the new transmission line (compacted and modernized) and to the Rail Auxiliary Line, and contains equipment demanded by the region (Madureira and adjacent districts), such as volleyball and basketball courts (strong presence in the neighborhood and social character), futsal, skateboard, children's toys, exercising equipment, orchard, artificial waterfall and beach (another recreational option, since the neighborhood is far from the sea), public lan house (Nave do Conhecimento Compositor Silas de Oliveira) and a space for cultural presentations (Teatro Arena Carioca Fernando Torres). However, as the greatest representation of the local community, the project has responded with the implementation of kiosks (a new version of the typically suburban bar) and a large stage (Praça do Samba) that gathers the samba circles and presentations of the samba schools from the neighborhood (Grêmios Recreativos Escolas de Samba Império Serrano and Portela). These are some of the social practices that are part of the daily life of its population.

On the other hand, although there are not many qualified spaces, the social practices have adapted themselves to the available spaces and, nowadays, have become cultural references. There are also areas such as those of the remains of the viaduct construction, which have become a problem, since they have been occupied by homeless people, drug addicts, or other marginalized groups in Madureira, and present an intense and peculiar appropriation of the neighborhood. The Baile Charme of the Madureira Viaduct (Negrão de Lima/Espaço Cultural Rio Charme) that takes place on Saturday nights, is a manifestation of the black culture of Rio de Janeiro, that expresses itself through musical rhythms such as charm and hip hop, since the 1990s. Therefore, besides occupying this space under the viaduct, is also represents an essential instrument in the diffusion of the neighborhood culture. Under this same viaduct, there is yet another form of appropriation of the space that occurs in an adapted and somewhat disorganized manner: the occupation by kiosks of fast foods and drinks (Fig. 2). They represent the of the traditional boteco (small bar) and operates in varied and different schedules conferring a diversified dynamic and vitality to the place.

Therefore, the centrality of the Madureira neighborhood reflects the reproduction of its everyday life, which was analyzed in the light of the relationship between agents-public spaces, producing local identities, and contributing to the update of the concept of centrality, which is now also based on elements of qualitative and symbolic order, rather than on economic indexes. These centralities, its population and users reveal what we denominate as multiple identities, those that since their origins have been fundamental to the constitution of the local identity.

Figure 2. Appropriation of the public space: Area under the Viaduct Negrao de Lima occupied by kiosks that operate at different times. Source: Author's Personal Collection, Nov 02, 2015.

ii. **Diversity of accesses:** Madureira represents an important nodal point in the city. It reflects not only upon its suburban neighborhoods and the West Zone, but also on cities of the Baixada Fluminese, such as Nova Iguaçu, Duque de Caxias, São João de Meriti, Belford Roxo and Mesquita. The neighborhood has two railroad tracks that physically demarcate the division of the territory into three parts (the side of the Campinho district, the side of the Meio and the side of the Mercadão de Madureira). The diversity of accesses is based on these roads, important components of the city's public transportation system and its surroundings: the first and oldest is the Central Railroad of Brazil (former Dom Pedro II Railway) – its first stretch was built between the parishes of Santana and Queimados, and the Madureira Station was only inaugurated in 1890 (Abreu, 2006, p. 50). Three branches go through this station today: Ramal Deodoro, composed by 19 stations in the city of Rio de Janeiro; Santa Cruz that goes through 23 stations, all inside the city; and the Ramal Japeri that goes through nine stations in the city of Rio de Janeiro and through 13 stations in cities of the Baixada Fluminense. The second railway line is the Auxiliary Line, with a first stretch inaugurated in 1893, linking Mangueira to Sapopemba, had its station (Magno/Mercadão de Madureira) inaugurated in 1898 (Abreu, 2006, p. 53). This station currently has an extension, Belford Roxo, which goes through 15 stations in the city of Rio de Janeiro and five stations in the Baixada Fluminense. The railway stations located in the neighborhood are important points for the modal shift because it allows passengers from the cities of the Baixada Fluminense to access other modals, such as buses to other parts of the city. It is also worth noting that, besides the possibility of changing branches at the Madureira Station (transfer station free of charge), there are other possibilities of modal shift, such as the integration with BRT (Vila Militar, Magalhães Bastos, Paciência and Santa Cruz and the Madureira Station itself, with fare). Another possibility of modal shift from the Madureira Station is the subway (Central do Brasil, São Cristóvão and Maracanã stations, with fare). There are also possibilities of modal shifts parting from the Magno/Mercadão Station in Madureira: the shift can be made to the subway (Central do Brasil, São Cristóvão, Maracanã, Triagem and Pavuna) and to the BRT, in a great diversity of itineraries, modal shifts, connections between points within the city, in the distant or close regions, trips between Rio de Janeiro and Baixada Fluminense, branch shifts (Fig. 3).

It also gives us a notion of the passenger flow going through Madureira and that can actually translate into the diversity of people, cultures, relationships and ways of appropriation and perception of spaces, that is, urban vitality.

When we analyze the bus modal, based on a survey divided by sectors previously demarcated by the preexisting fragmentation in the neighborhood center due to the railroad passage, this flow is further amplified (Fig. 1):

Figure 3. Diversity of transportation modes: BRT station (on the right), access to Madureira Railway Station (on the left) and bus stop (at the bottom). Source: Author's Personal Collection, Nov 02, 2015.

a. in the "Campinho side" there is an end point for three bus lines to São Gonçalo (Alcântara and Tribobó), besides other stops where six lines connecting different regions of the city go through, and four more lines with connecting points from Rio de Janeiro to the cities in the Baixada Fluminense (Nilópolis and Nova Iguaçu);
b. in the "Middle side" there is an end point for 11 bus lines to various points of the city and five to the Baixada Fluminense (São João de Meriti, Belford Roxo and Nova Iguaçu), 13 lines connecting different points in the city have bus stops in that area, 14 lines of the Baixada Fluminense (Duque de Caxias, São João de Meriti, Nova Iguaçu and Nilópolis), destined to other points of Rio de Janeiro, and two intercity bus lines, with lines to Petrópolis and Santa Cruz da Serra;
c. in the "Mercadão side" there is a terminal for four bus lines to various points of the city and five to the Baixada Fluminense (São João de Meriti and Nova Iguaçu) and São Gonçalo (Alcântara and Tribobó); three lines with connecting points in the city have stops in this area, and 12 lines from the Baixada Fluminense (Duque de Caxias, São João de Meriti, Nova Iguaçu, Belford Roxo and Nilópolis) that stop at another point in the city.

Therefore, the centrality of the Madureira neighborhood is evidenced by the accessibility and degree of influence, which was expanded with the implementation of the expressway of the Transcarioca BRT, the construction of the bus stations, Madureira and Mercadão, and the renovation and expansion of the Paulo da Portela bus terminal. This ideia is further reiterated by the fact that the neighborhood has another nine lines, connecting the end points of the express corridor (Tom Jobim Airport and Barra da Tijuca) besides other stops that offer access to several other neighborhoods.

iii. **Diversity of activities and uses:** The Madureira neighborhood has a commercial and service reference position that goes beyond the municipal limits, reaching also the neighboring cities of the Baixada Fluminense, such as Nova Iguaçu, Duque de Caxias, São João de Meriti, Belford Roxo and Mosque. Firstly, commercial activities that are in a formality will be treated: they are divided into large anchor stores that belong to national and international networks (Casas Bahia, Ponto Frio, Lojas Americanas, C & A, Marisa, Supermarket Champion, Casa & Video, Insinuante, Ricardo Eletro, Supermarket Super Network, Leader, Supermercado Multi Market etc) or in commercial buildings—ground floor used by one or more stores, and the other floors that are divided into commercial rooms—and also the buildings of mixed use—ground floor for trade; and the other floors shelter the housing use. These establishments are spread over several streets, concentrated in the commercial center and in the main commercial streets, such as Domingos Lopes Street, Portela Road, Avenida Ministro Edgar Romero (both sections), Rua Maria de Freitas, Travessa Almerinda Freitas, Rua Carolina Machado and Rua João Vicente, but also in others more widely. It can be seen that there is a significant variety of types of products traded in retail, such as household appliances, residential and commercial furniture (modulated, planned or loose), clothing in general, footwear, bedding, table and bathroom textiles, household utilities, products for personal use, products for handicrafts, disposables and packaging, food in general (snack bars, restaurants, food stores, butchers, groceries, sweets etc), drugstores, pharmacies, bookstores and stationery, orthopedic products, optics, mobile telephony, hairdressing products, others for parties and decorations etc. (Fig. 4). It was also verified that some establishments have a branch in the Madureira Shopping Mall, or in the Condomínio do Entreposto Mercado of Rio de Janeiro (Mercadão de Madureira), which in recent decades has been directed towards the concentration of products for parties and decorations, sweets and religious items.

Madureira is also known as a place of popular and informal commerce occupying several areas of the neighborhood center. In some situations this commerce is carried out in an organized and standardized way in covered places (Rua Carolina Machado nº 534, Rua Soares Caldeira); or in an agglomeration of standardized stands in open areas (next to the Silas de Oliveira Viaduct—BRT Transcarioca—at the confluence of Conselheiro Galvão Streets and Avenida Edgar Romero and access to the footbridge to the Mercadão de Madureira /Ramal Linha Auxiliar); or in standardized but disorganized stands (under the Viaduct Negrao de Lima, former camelódromo of Rua Francisco Batista, which was destroyed by the constrction of the Silas de Oliveira Viaduct); or even as a concentration of street vendors in a very improvised way, with no stands, in uncovered places (corner between the Avenida Ministro Edgard Romero and Estrada da Portela, next to the access to the Magno/Mercadão Railway Station of Madureira and in the accesses to the Station Railroad Madureira/Deodoro Branch). By observing the pedestrians, it is possible to infer that they are not troubled by losing sections of the sidewalks, ramps, stairs and catwalks to the street vendors. On the contrary, they take advantage to make quick purchases of various products (batteries, socks, cell phone chargers, swabs, toothbrushes etc), biscuits, soft drinks and water. From this perspective, it is believed that the presence of street vendors, organized or not, standardized or not, reinforce the characterization of Madureira as a place of commerce and central to specific identity elements, even though the need to produce income by these social groups was already established.

In terms of service provision, we have located establishments that offer maintenance and technical assistance for computers and appliances, training courses for drivers, gyms (small and large networks such as Formula, Curves and Smart Fit), medical clinics (general practitioners, orthopedics, ophthalmology, gynecology, pediatrics, cardiology, gastroenterology, etc.), clinics in general (physical therapy, nutritionists, psychology, therapy, radiology exams, etc.), clinical analysis laboratories, medical attendance clinics (CTO Clínica Traumato-ortopedia e Casa de Saúde e Maternidade Madureira), language courses (Wizard, Wise Up and Ibeu), computer courses (Microlins, Microcamp and Microrio), colleges (Curso Martins, Pereira Lopes, Pensi Colégio e Curso, Sistema Elite de Ensino, Liliane Travassos, Santa Mônica Centro

Figure 4. Edgar Romero Avenue: Diversity of activities and great populational fluxes. Source: Author's Personal Collection, Nov 02, 2015.

Educacional and EMEC Educacional), professional training courses (Curso Souza Barros de Enfermagem, Escola Técnica Electra, Escola Técnica ETEP RJ e Senac), banking and financial services (three agencies of Banco Bradesco, three branches of Banco do Brasil, three Banco Itaú, two Banco Santander, two Caixa Econômica Federal, two Banco BMG, one BRB Crédito Financeiro and one BV Financeira) and cinemas (Kinoplex Madureira of Madureira Shopping).

5 CONCLUSION

Considering the present discussion, it is possible to affirm that Madureira is configured as a space of symbolic representations and rich in meanings that fulfill the purpose of expressing social structures and translated by the spatiality through which identity gains materiality and produces urban vitality.

At the present time, when the debate about the city revolves around the discourse of globalization, especially regarding the redefinition of the roles cities play and their competitiveness, and when there is an increasing homogenization carried out through the production of global landscapes, local identity emerges as a fundamental form of resistance in face of the to the primacy of the globalized urban scenario.

According to Serpa (2011), in this context it is difficult to talk about of "places" as spaces of experience. That is, places that in some way are able to subvert or question the hegemonic logic of the production of urban spaces in capitalist metropolises. It seems, however, that such "places" do exist and persist inside the metropolitan "gaps", especially in their more popular areas. There is a well-known affirmation from this author in which he states that "in the popular neighborhoods of capitalist metropolises, the inhabitants are the true agents of transformation of space. They articulate themselves in "network", not in a single network, but in overlapping networks. (...) demonstrating that there are multiple representations of these spaces, among the groups/agents that compose their networks of social relations. It is then revealed that neighborhoods are transverse cultures that encompass many and many subcultures" (Serpa, 2011, p. 98).

In conclusion, the path undertaken in this essay reaffirms the production of another city that is not a scenario, but is authentic and reinforces the presence of a local identity, as well as represents resistance (although unintentional) to the homogenizing process legitimized by the enterprise of large urban projects as a response to a globalized capitalist economy. We have realized that the conclusions reached in this work were only possible through the immersion of the researcher in the daily life of the place, as a participant observer, looking at how people appropriate and relate to the spaces, and that social and cultural practices are lived in the public space. There is nothing more democratic and representative for what the concept comprehends.

ACKNOWLEDGEMENTS

Our sincere thanks to CAPES—Coordination for the Improvement of Higher Education Personnel, for the promotion of scholarship (Program Doctoral Sandwich Abroad—PDSE 2016, Process n° 88881.135287/2016-01; Faculty of Architecture, University of Lisbon), allowing the complementation of this research.

REFERENCES

Abreu, M. de A. 2006. A evolução urbana do Rio de Janeiro. 4nd edn. Rio de Janeiro: IPP.

Bernardes, L.M.C. 1990. Rio de Janeiro: cidade e região. Rio de Janeiro: Secretaria Municipal de Cultura/Departamento Geral de Doc. e Inf. Cultural.

Corrêa, R.L. 1997 – Trajetórias geográficas. Rio de Janeiro: Bertrand Brasil.

Gehl, J. 2013. Cidades para pessoas. 2nd edn. São Paulo: Perspectiva.

Jacobs, J. 2000. Morte e vida de grandes Cidades. 1nd edn. São Paulo: Martins Fontes.

Lynch, K. 2007. A boa forma da cidade. Lisboa: Edições 70.

Serpa, A. 2011. 'Lugar e centralidade em um contexto metropolitano'. In. Carlos, A.F.A., Souza, M.L. de, Sposito, M.E.B. (ed.) A.

Drawing (...) City (...) Body, Dwelling on Earth – Janeiro (Ed.)
 ISBN 978-0-367-02602-8

City Portrait: A pedagogical experience in the Faculty of Architecture and Urbanism of University of São Paulo

L. Jorge
Faculty of Architecture and Urbanism of University of São Paulo, Brazil

ABSTRACT: The paper introduces the classic and known problem about the architecture's teaching: the urban text's interpretation or, in another terms, the understanding of the city as a language's phenomenon. Its focus is in methodological approach, detaching the questions about the expressive ways or the instruments of representation of the city. Using the body to perceive and to know the São Paulo city (Brasil) to produce information and to discover how to organize them in narrative structures, the students did these works in their last year at Faculty of Architecture and Urbanism of University of São Paulo (FAU-USP), as Undergraduate Final Project. With some examples to illustrate the main questions involved in the relationship between interpretation and creative production of the city's analyses (almost all of them, done to visual ways), the paper focus the pedagogical practice in the architect's education.

Keywords: City's reading, representation, body, strolling, pedagogical practice

1 INTRODUCTION

La vida no es la que uno vivió,
sino la que uno recuerda y
cómo la recuerda para contarla.
Gabriel García Márquez

There is a fascination with the city of São Paulo in the students of FAU-USP that challenges them to face it as multifaceted, polymorph and complex knowledge object. Everyone see in the city an enigma that resists to the objectivity of the analytical instruments from the history of the city, the social sciences, the political economy, the urban planning and the urbanism. Hovering over the structural interpretations of major scientific consistency, there is a dense layer of meanings inviting the deciphering, the reading and the perception of other unreachable aspects, like the singularities of places or of everything that makes São Paulo unique. Or multiple. This layer or orbit of problems, in general, do not contrasts to the referred "scientific" analysis: it only indicates, to those students, a necessity of saying more or in another way, given the feeling of insufficiency that remains after the most structured visions by the disciplinary knowledge and, because of that, more general. The problem of the city identity is being imposed, naturally, by a constellation of related themes: the memory, the micro-history with its anonymous characters, the picturesque or the curious emerging observation of the streets, the notable and acute artistic register and the affections, the emotional narratives and the means by which they are carried out, the practice of coexistence and the life in the public spaces, the walk through the city and the drifts, everything that we can find in the cultural field.

The diversity—with every meaning that this word may carry—and the scale of the city of São Paulo afflict the students in imposing their epistemological obstacles, its complexity e the inherent difficulties to perceive an expression of totality. However, both characteristics also fascinate and motivate them to look for other forms of reading appropriated to the architect and the urbanist in training at FAU: other perspectives, other sayings or other portraits that no longer seek a resolution or a final meaning to this inexhaustible and unmanageable reality, but, deliberately, seeks to seize it as poetic matter, as an infinite source of inspiration and creative motivation.

The students venture through the numerous São Paulo, traversing territories and going through collected topics in the practice of space, tracing sentimental or imaginary maps, registering, through images, the raw material to their respective investigations about the city and about their own registers, at the same time. The election of the expressive ways or the instruments of representation of the city launches a reflection field as important as the object to which they refer: to use familiar means of register or take advantage of the opportunity to also investigate the characteristics of these means, the possible dialogues between them until the reading is found, in elaboration, with the forms as it manifests or realizes in the intrinsic properties to these own means. In this develop of work arise the domain of the discourse and the personality of the author, when form and subject can't be separated anymore, without the loss of the meaning. To achieve this authorial awareness, only after lots of tries and expressive experiences, taking risks and dealing with scarce time. The risk is to not elect, previously, a mean of representation (the easiest way) to portrait the city that seeks to reveal, but find it in the reading activity, in the expressive experiences triggered to caught the reality in focus, to become, indeed, a "thinking portrait", subject and drawings, undivided.

FAU cultivates an environment that stimulates this kind of work, the approach to city by images, the non-verbal narratives, with the most distinctive means: drawing, photography, video, collage, animation, painting, engraving, poetry. These ways to look to the city were cultivated continually in the diversity of the theoretical and practical disciplines of the course and in each challenge of elaboration of a project, when the city is a problem permanently renewed. Because of that, it is natural, at the ending of the course, that the students' attention is focused on the city where they graduated, properly accompanied by an creative, expressive and discursive will. They want to talk, but do not know very well about what. Or they want

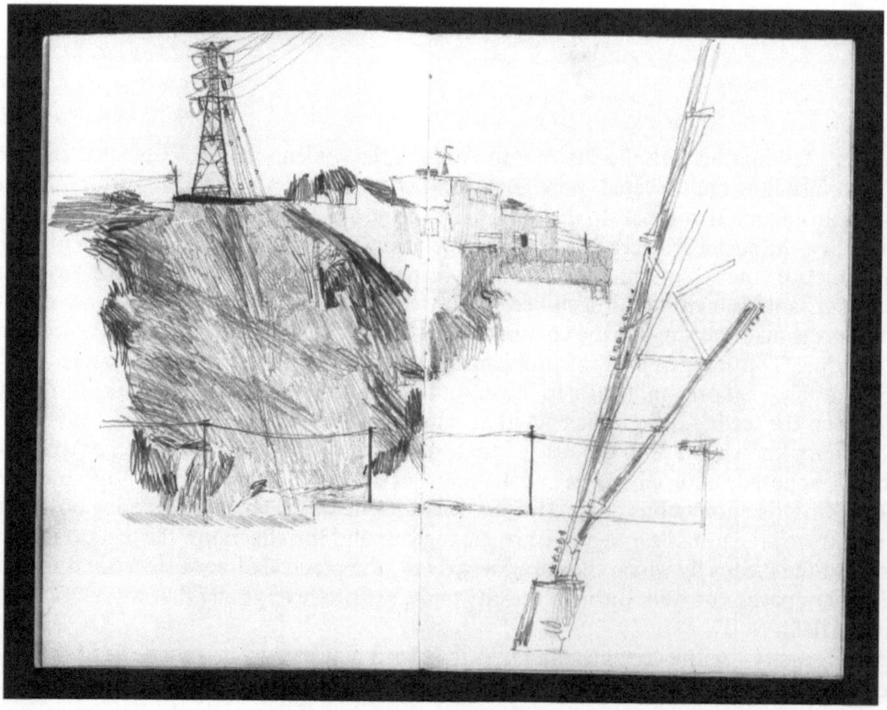

Figure 1. Andrés Sandoval (Blank Space, 1999).

to talk about something, but do not know very well how. The beauty in this initial power is in the lives that move, that are amazed and surprised by inquiring themselves. Acknowledging, as well observed by my spiritual guide, Paul Valéry, that the thought is, in short, the work that originates in us something does not exists.

The city that presents itself full of gaps, spaces without information, forgotten or unfinished, compose the drawings done in a sketchbook for the field work: walk through the city to find the object of the drawing in loco. An exhbition of 50 chosen drawings of various produced sketchbooks was the form of the presentation of the student Andrés Sandoval, in this Undergraduate Final Project (UFP), called Blank Space (1999). We are taken by a route made with minimum recognizable elements—allusion to São Paulo—between various anonymous referents of the revealed view.

Walking through a banal and ordinary São Paulo; through its spaces of abandonment or unfinish, without the signs that identify it and in which we recognize ourselves as participants. The traces of the banal and big city, where the view of chaos reigns, where the information is not organized in known addresses or is not fixed in memorable views. To walk looking for a senseless view, deprived of recognition and identification, like it is only a visual structure, so that the act of drawing it can be analogue to a primordial writing, producer of a meaning not imaginable before. As if the drawing of the urban view—so worn that is no longer noticed—could reveal a second perspective, a zero degree of the language that inaugurates.

The UFP of Julia Sayeg Tranchesi (***Drawing Route Landscape***, 2014) brings an analogy between the walking and the drawing. Moved by the motivation to walk long distances and to distance herself from the known zones of the city, the student registered what claim her attention in those long routes, taking into account a serial vision, as Gordon Cullen. In the development of the work, the sketchbook's register indicate themes to be studied more accurately in another scale, transposed to paper of great dimensions to be filled and surpassed, analogously to the long routes, by the filling of their empty and blank spaces and by their density of the graffiti.

The will of becoming a traveler through São Paulo is as common as attractive for the students. It is worth remembering that Eclea Bosi distinguished two types of narrators: the foreigner that narrates his adventures and travels, and the resident or native that knows his lands, his people and the past that surrounds them. These students are like a mix of both these types of narrators: they are adventurous people of São Paulo, with long distances to walk and places to know in the city, at the same time close and distant. Everything in order to add pieces to a puzzle in eternal assembling, to find meanings that scape of their own living experiences in this city. Walking to narrate, paraphrasing the title of the memories of the writer Gabriel García Márquez (*Vivir para contarla*). Rambling, loitering, toddling, traversing, drifting are used verbs, taken from the writings of authors such as Poe, Baudelaire, Smithson, I. Calvino, G. Simmel, W. Benjamin, G. Debord, R. Sennet, F. Careri, P. Bereinstein. To pursue the form of registering this practice is imposed since the first travel.

Figure 2. Julia Sayeg Tranchesi (Drawing Route Landscape, 2014).

Figures 3 and 4. Isabel Falleiros Nunes (São Paulo by feet: stroll and passage, 2007).

São Paulo by feet: stroll and passage (2007) de Isabel Falleiros Nunes is one of these works, inspired by the concept of "drift", proposed by Gui Debord, where the practice of incursion in the city spaces was analogycal to the search for the way to map it or make a representation of the spaces travelled as a narrative field. To catch the dynamic of flows, paths were redone in distinct times: dawn, the rushed way to work, the useful time of the day, the tired way back to home, the weekend. The city's pulsation: time and life in spaces.

Photography is a necessary language complementary to the formation of the architect. At FAU, many put it in first place in professional intentions. The ones who dislocate it from the obvious condition of "work instrument" or "language-media" to the place of elected aesthetic and intellectual motivations, to the final work, the synthesis of all the work. Professor Arlindo Machado's excellent book (*A Ilusão Especular*) have always accompanied me in the orientations and classes when the photography is mentioned. According to my judgement, the best book about photography ever written (yes, better than the great approaches of R. Barthes, S. Sontag, W. Benjamin, A. Bazin or B. Kossoy). I also stimulate the students to know the works of classic photographers: Robert Frank, William Klein, Louis Stettner, Walker Evans, Andreas Feininger, Henri Cartier-Bresson, Robert Doisneau, André Kertész, Raymond Depardon, Tore Johnson, Harry Callahan, Diane Arbus, Reinhart Wolf, Keizo Kitajima, Matthew Klein, Luigi Ghirri. And the ones "from Brasil": Sebastião Salgado, Maureen Bisilliat, Pierre Verger, David Drew Zingg, Anna Mariani, Bob Wolfenson, Cristiano Mascaro, Mario Cravo Neto, Thomaz Farkaz, Araquém Alcântara, Marcel Gautherot, Claudio Edinger, Miguel Rio Branco, Geraldo de Barros, Arnaldo Pappalardo, Antonio Saggese, Ed Viggiani. Invariably, I end up knowing the works of younger photographers brought up by the students themselves, much more attentive to the present of art, like, for example, the priceless *Cartiê Bressão–liberte, egalité et brasilité* (2013), pseudonym and title of Pedro Garcia de Moura's book.

São Paulo hand-to-hand (2010), UFP by Inês Pereira Coelho Bonduki, was developed in the drift practice, when the errant body of the author was going through large portions of the city's body, having the body of others as focus of the look. Bodies in everyday action, in the urban space practice, in the frequency of places, submitted to every sort of necessary movements to act or live in São Paulo. Walking long nocturnal distances to reach a mean of transport, compressing yourself to enter it, squirming yourself to support the endless waiting standing up. To catch the body sitting down after exhaustion, supporting itself for a pause, distended after the compression, hanged, running, jumping, leaping, going down, soaring, flying, dancing—with photographic images is one of the ways of critical thinking of the architect and urbanist about the conditions of the hard life in the city, about its broad-spectrum contradictions, so deep and delicate.

In *Window, photography and city* (2012), Julia Paccola Ferreia Nogueira establishes syntactic relations between the framework of photography and the windows of architecture or the made up in the urban landscape by the photographic act itself. Analogies between the two forms of writing and spelling with light complement a theoretical and practical reflection in which the perception of the city and the photographic language complement each other like two faces of a thinking or a knowledge established in the field of metalanguage.

Figure 5. Inês Pereira Coelho Bonduki (São Paulo: hand-to-hand, 2010).

Figure 6. Julia Paccola Ferreia Nogueira (Window, photography and city, 2012).

As a similar starting point but with a phenomenological approach about the problematization of perception, *Lights and senses: from the city to the screen* (2015) by Aruã Oliveira Wagner, is the result of the reflections and walks around the city, at night, to photograph the brightness of the facades formed by randomly illuminated windows. Those images offer the visual raw material translated to another mean and technique—serigraphy—where the field of photo-chromatic research is established to portray the noticed urban landscape.

In *Passages in the city* (2013), Luisa Magni Fecchio starts from the classic and namesake text by the philosopher Walter Benjamin about the Paris of the 19th century to guide her in São Paulo, searching for situations and paths that are rich in vitality and urbanity. Going

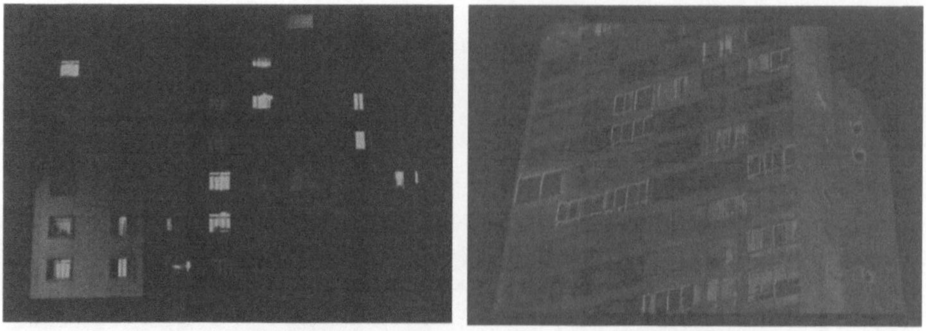

Figure 7. Aruã Oliveira Wagner (Lights and senses: from the city to the screen 2015).

Figure 8. Luisa Magni Fecchio (Passages in the city, 2013).

through the galleries installed in buildings of great size and many uses of the "New Town" of the city, between Vale do Anhangabaú, Praça da República, Largo Paissandu and COPAN building, the showcases are noticed as privileged signs of the gallery space and its commercial and social activities, but above all as a microcosm of the diversity of urban universe, formed by a constellation of characters in their little work spaces. In each showcase, a frame of life in the city, the metonymical expression of its multiple faces and meanings. The frontal framework of each showcase, the use of black and white, the record of artificial light and the human presence constitute the pictures of a life in course.

Correspondence Project (2013) by Manuela Costa Lima is a work of walk and errancy by cities all over the world, through a virtual drift along the streets seen by Google Street View. The result is a work I called "re-photography", which means the process of selecting and clipping scenes of this great visual flow offered by the app. The interest was to photograph people who appear in these images to look for a communication with them, through the dispatch of physical letters, by the conventional system operated by post offices. More than five hundred letters were sent to the most diverse places of the world, with the re-photographs

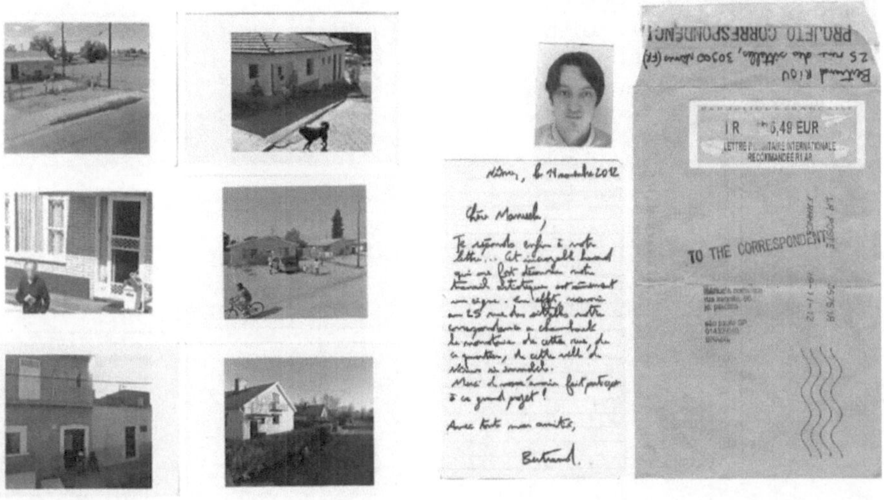

Figure 9. Manuela Costa Lima (Correspondence Project, 2013).

printed and an extra envelope so that the people could answer, through the same mean, whatever they want to from the suggestions in the letter, asking who the recipient is and about the work proposed: to correspond and find character of life, so multiple and variated, so far and so close, in the traffic between the virtual and the material world. In the beginning, the chance was the procedure that led to the drift. Slowly, the practice of drift taught how to get to special places—in the most broad and variated meanings as possible—and to fascination that the diversity of reality is capable of offering insinuating imaginary narratives about the banality of places, about the human condition in the urbanized world and amplified by the technologies of digital communication. Like bottles thrown at the sea, many letters received no answer, many letters didn't find the supposed address of the hypothetical recipients photographed in front of a house that was imagined to be theirs, but, amazingly, some letters were responded. Life stories: places.

The mirror of the metropolis (2015) by Ana Cristina Niessner is a demonstration that every elaboration of interpretative hypothesis about the urban space comes from a dialogue between the direct experience with the city and the reflexive and aesthetic motivations originated from such diverse sources like philosophy, history, theory, art or literature. The mirror of the city here caught, refracts more than reflects the acknowledged information. Acknowledging the studies of E.H. Gombrich and the confirmation that the meaning of order is intrinsic to the elaboration of a language, to the communication of a meaning and the construction of a narrative, the work dodges from an unequivocal and finalist speech and to be only search, plural actions travelling through multiple ways of representation (analogical and digital) to revisit the classic theme of pattern, repetition, of order in the contemporary cultural context.

Baronesa (2008), UFP by Claudia Afonso is a video that portrays building "Baronesa of Arary" residents' stories, in Paulista Avenue, São Paulo City. It is a work of extreme delicateness and emotion, in the exposure of content—a kaleidoscope of lives in the metropolis—and in the denudation of the narrative form, in the simplicity of the treatment of images, in the edition of phrases, composing a mosaic of impressions and meaning of the city. Now, about the habitational use, a portray of a microcosm that contains São Paulo's diversity.

Light: body-space and black holes (2012) by Adriana Maiolini Mesquita is a video recorded at the ruined square of the old Bus Station of Luz Neighborhood. Heart of "Crackland" and the conflicts between the population and resident uses and the new urbanistic order proposed to the "New Luz", the referred area and its surroundings transform in stage and scenery to an unusual dance: a plastic-poetic action led by quartet of ballerinas who appropriate banal

Figure 10. Ana Cristina Niessner (The mirror of the metropolis, 2015).

urban elements to introduce the extraordinary in chaos, history, the culture of the place and its old and current habitants in a critical manifest to the project and it *modus operandi*. The sound and image edition visits popular tradition, the "*Boca do Lixo*" cinema and the cruel contemporary reality, in artistic key, by the fusion between performance and cinema, between body and city.

Invariably and constantly, I recognize in the orientation of these works voices of other professors and of other disciplines, constituting a silent dialogue between my orientation and

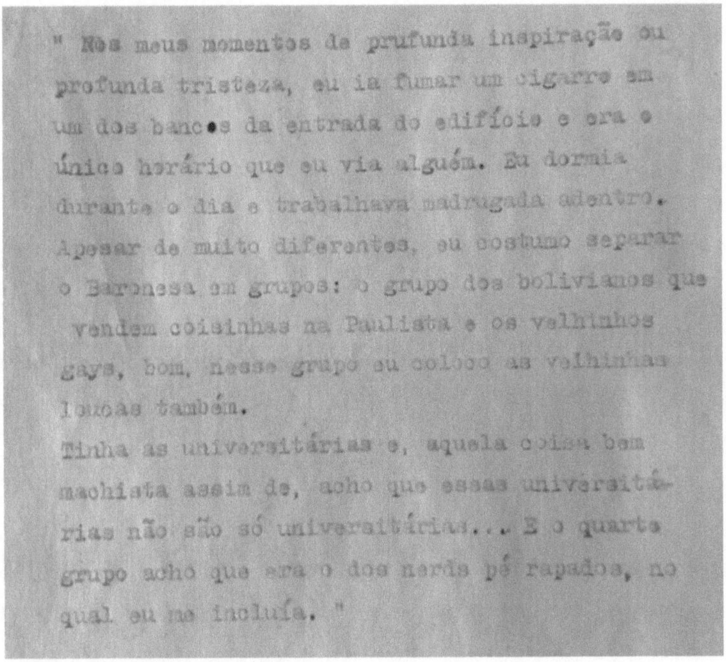

" Nos meus momentos de prufunda inspiração ou profunda tristeza, eu ia fumar um cigarro em um dos bancos da entrada do edifício e era o único horário que eu via alguém. Eu dormia durante o dia e trabalhava madrugada adentro.
Apesar de muito diferentes, eu costumo separar o Baronesa em grupos: o grupo dos bolivianos que vendem coisinhas na Paulista e os velhinhos gays, bom, nesse grupo eu coloco as velhinhas loucas também.
Tinha as universitárias e, aquela coisa bem machista assim de, acho que essas universitárias não são só universitárias... E o quarto grupo acho que era o dos nerds pé rapados, no qual eu me incluía. "

Figure 11. Claudia Afonso (Baronesa, 2008).

Figure 12. Adriana Maiolini Mesquita (Light: body-space and black holes, 2012).

what I can realize FAU meant to each student until that moment. According to this perspective, much individualized by each student's life experience, I never listen to the same FAU, as it is never presented the same São Paulo. But the "verb-vocal-visuality" of this omnivorous "sãopaulistness" city has always been and always will be object of stupefaction, amaze and enchantment of these innumerous portraitist architects who graduated themselves, graduating the FAU.

REFERENCE

Jorge, L.A. 2016. A teia e o rio. Tese de Livre Docência, Volume I. São Paulo: FAU_USP.

Drawing (...) City (...) Body, Dwelling on Earth – Janeiro (Ed.)
© 2019 Taylor & Francis Group, London, ISBN 978-0-367-02602-8

Legal alien, exile and intermittent homes in "Englishman in New York" (Sting, 1987)

L. Sol
CIAUD, Lisbon School of Architecture, Universidade Lisboa, Lisbon, Portugal

ABSTRACT: "Much of the exile's life is taken up with compensating for disorienting loss by creating a new world to rule. [...] The exile's new world, logically enough, is unnatural and its unreality resembles fiction" (Said, 2000: 144).

Taking into account the notion of Exile by Edward Said (Reflexions on Exile, 2000) and that of Emanuel Levinas's Dwelling (Totality and Infinity, 1979), the city of New York will be examined through the video clip "Englishman in New York" (Sting, 1987), directed by David Fincher.

New York has been from its beginnings a city of foreigners. The foreigner is the one who establishes himself in the place where he arrives with the memory and the habits of the place behind. It is through the look of a stranger and based on this conflict (from the place that is left versus the place where one arrives). This text will approach the territory that will henceforth be the Home—bearing in mind its ephemerality and volubility—of this Englishman in New York.

Keywords: New York, Exile, Dwelling, Alien, Home

1 INTRODUCTION

In his book "Relections on Exile", Edward Said (2000, p.148) points out that seeing "the entire world as a foreing land" enables one to a particular originality in the gaze in that most people are aware of a culture, a place and a home. For Said, the notion of exile is mainly intellectual, which makes it possible to be exiled in his own country.

New York has always been a city of arrivals and departures, in which a foreigner has always arrived with the aim to start over. It is precisely on the look of the forigner that the music video of "Englishman In New York" (David Fincher, 1987) focuses on.

An exiled individual has at least two homelands allowing plural visions, simultaneous dimensions, and an awareness of what Edward Said has called "counterpoint." By "counterpoint" Said means the permanent conflict between the habits, expressions and activities of a place where one arrives and the memory of those same habits of the place that one has left: "I Don't drink coffee I take Tea May dear/I like my toast done on one side".

Sting's New York, by the eye of David Fincher, begins by being filmed syncopately in glimpses, between moving and still images. The Upper West and Upper East Side are viewed from Central Park, the Fifth Avenue, Greenwich Village, the New York Stock Exchange Building on Wall Street, the snow-capped Central Park, the Public Theater, Tyffany's entrance, and a coffee shop. The city—New York, Manhattan—is filmed in black-and-white, foggy, nostalgic and in an atmosphere of melancholic bustle.

Sting, who walks under an umbrella in the streets of Manhattan, gives voice and personifies a wandering, that, under an anonymity that is always about to be denounced, contemplates and looks at what the new world has of strange and familiar – "You can hear it in my accent when I talk/I'm an Englishman in New York."

The alien lies between the strange and the familiar, the proximity and the distance, between the place he left and the place where he has arrived. Both "the new and the old environments are vivid and actual, occuring together contrapuntually." (Said, 2000, p.148)

Sting wanders through the streets, he is English, he is in New York, but his role is mostly of a storyteller. The English poet, writer, performer, actor and comedian Quentin Crisp (1908–1999) appears simultaneously to the verse "See me walking down Fifth Avenue/A walking cane here at my side". He is the metaphor of "Englishman in New York", it is about him and the City of New York that the music video focuses, and it was for him that Sting wrote the lyrics of this theme.

Quentin Crisp was born in England and self-defined as an exhibitionist and a subversive homosexual. In the twenties and thirties he wandered the streets of a conservative and homophobic London wearing eccentric dresses, painted nails and long hair, trimmed and dyed with unusual colors.

In this age, this audacity was rewarded with violence and humiliation. He was judged for sexual harassment where he was a self-advocate – "Perhaps my very existence is the form of importuning. Otherwise I am not guilty as charged" (The Naked Civil Servant, 1975) – where he was acquitted, attempted to serve in the army at the outbreak of World War II (he was rejected as being clinically declared a "sexual pervert"), and in 1981, at seventy-five years of age, he moved to New York.

> "The moment I saw Manhattan I wanted it! It was more like the movies than I had ever dreamed. But did Manhattan want me?" (Englishman in New York, 2009).

The concept of exile, being related to territory, constitutes, first and foremost, a state of mind that implies the adoption of a critical distance in relation to all cultural identities and a permanent opposition to all orthodoxies, as well: "There is also a particular sense of achievement in acting as if one were at home wherever one happens to be." (Said, 2000, p.148).

Quentin Crisp appears with a staggering pace with a hat, gloves and an overcoat. He embodies someone who has always been a foreigner, both in his own land and in foreign

Figure 1. "Englishman in New York", Sting, 1987.

lands. In England, where he was born and lived until he was seventy-five, Crisp was always a "stranger." Strangeness diffuses here with a kind of deterritorialization that is intrinsically related to a sense of insecurity. In the "Naked Civil Servant", Crisp says in an exalted tone to a friend: "You live in a secure world of normality." The purpose of this statement is not to affirm the normality of others but, most importantly, to underline his own misfit and consequently, the threats that this poses to his own safety.

The sense of insecurity and hostility are symptoms of someone who does not feel at home: "Borders and barriers, which enclose us within the safety of familiar territory, can also become prisons." (Said, 2000, p.147) Exile entails, not only, a displacement into an unknown territory, but also an escape/rupture with the place that was left. The limits that define the boundaries of affection and belonging are volatile and ephemeral. Said (2000, p.147) points out, that "what is true of all exile is not that home and love of home are lost, but that is inherent in the very existence of both." The space chosen by a foreigner to be sheltered, for this reason, it is first and foremost, a space of fiction. Constantly staging himself, the fiction space choosen by Quentin Crisp, is also a place where he did not feel successful in his own role,

> "When these people came to me and said: 'We should like to make a film of your life' I said 'Yes, do. Films are fantasies. Films are magically illusions. You can make my life a fantasy as I have tryied but failed to make it.' But then they said they wanted the film to be real, you know, real life. So I said 'Any film, even the worst, is at least better than real life'. Then they said, 'well of course we should have an actor to play you' I said 'I spent sixty-six years on this Earth, painfully attempting to play the part of Quentin Crisp. I have not succeeded. Yes of course you must have an actor to play me. He will do it far better than I have done." (The Naked Civil Servant, 1975).

The place where the wanderer moves around is this platform of reconfiguration, reinventing the individual, making it possible to start his life again. Displacement allows us to

Figure 2. "Englishman in New York", Sting, 1987.

understand our existence in this society that constitutes a false shelter whose real purpose is to camouflage the violence, the lack of meaning, the solitude and the chaos that proliferate in it. Security is a state of mind. The itinerancy between one context and another, allows us in one hand, to escape and avoid the threats of a place and in the other hand, to face the obstacles that are in the destination:

> "When I arrived in America, with all my worldly goods tied up in a red handkerchief, I was betrayed. In 1980, I had met a man who had said I could stay with him indefinitely when I arrived. I wrote to him from England telling him of the day and almost the hour of my arrival on the other side of the world. He had not replied so I had assumed that he was waiting for me. He was not. Indeed, he did not even open the door of his apartment but made me inaudible replies to my announcement of my presence, so I went to my only other friend in America." (Crisp, 1997: 05,06).

The ambivalence between what remains and what is to be found is the "counterpoint". There is a confrontation/overlapping of empathies and adversities of a place that we can call home. According to Levinas (1979, p.152) "the privileged role of the home does not consist in being the end of human activity but in being its condition and, in that sense its commencement." In other words, New York is the resumption of Quentin Crisp.

Crisp's realization when he arrives in Manhattan (Englishman in New York 2009) is that "It was more like the movies than I had ever dreamed of." Movies are intrinsically connoted with 'dream' and therefore with a 'promise of fulfillment'. To make a dream a reality, or an illusion as its own world is to some extent 'to become' and 'to be'. Meaning, the "first take from as a world, is accomplished as the home" (Levinas, p.152). The place that is called 'home' is also the place where one 'is'.

"The moment I saw Manhattan I wanted it!", the empathy demonstrated by Quentin Crisp (Englishman in New York 200) on his arrival to New York refers to a familiarization, a recollection and welcoming to this place. "Familiarity is an accomplishment [that] constituted [...] as dwelling and inhabitation" (Levinas, p.155,156), this affinity is described as follows:

> "The streets of New York are the closest thing to heaven on Earth. Here, you are not scurrying from one place to another as fast as your feet will carry you. You are on display. Without her outcasts, the metropolis would be a very dull place indeed. To succeed in heaven, you must avoid the shadows. If the sunny side of the street is full, walk in the road. And no account learn the language. The more English you sound, the more likely you are to be believed." (Englishman in New York 2009).

And therefore, "To exist hence-forth means to dwell" (Levinas, p.156). But once the address is found, Quentin Crisp (Englishman in New York 2009) asks "But did Manhattan want me?". Whilst London its "existence was a form of importurning" (The Naked Civil Servant, 1975), while in New York "You are on display" (Englishman in New York 2009). Crisp's words about New York have several allusions to how exuberance – instead of being seen as inadequate and offensive – is inclusive and even necessary – "To succeed in heaven, you must avoid the shadows" (Englishman in New York 2009). New York, has always been a city of arrivals and departures in which a foreigner always arrived, with the aim of starting over. Being a foreigner is, and always has been, an intrinsic characteristic of New York City. It is for this specific reason that the City has allowed Quentin Crisp, his "recollection", because " it is a recollection, a coming to oneself as in a land of refuge, which answers to a hospitality, an expectancy, a human welcome" (Levinas, p.156). Being in the land of asylum and this hospitality are the conditions that allow to" Be yourself/No matter what they say" to which Sting refers.

In "Englishman in New York" (Sting, 1986) the City is filmed in a syncopated manner, with fleeting, tremulous and cadenced images, where traffic images are alternated with images of street fronts. The outline of buildings between sky and snow, with roaming between the inside and the outside. Everything here seems to refer to the intermittent movement of the 'foreigner', since "Exile is life led outside usual order. It is nomadic, decentred, contrapuntal" (Said, p.149). This conflict of proximity-distance is also reflected, and above all, in the

Figure 3. "Englishman in New York", Sting, 1987.

statute of "Resident Alien" that was attributed to Crisp when he settled in New York. The fact of being allowed to reside in that country, but never granted citizenship, places him in the hybrid condition of "legal alien" where the reach of the "promised land" brings with it a simultaneous unattainability. In this way, the subject of "I'm an alien, I'm a legal Alien/I'm an Englishman in New York" confirm Said's thoughts (p.140) when he states that "exile [...] is fundamentally a discontinuous state of being" and where the notion of home is always a provisional one.

REFERENCES

Crisp, Q. 1997. The Resident Alien, London: Flamingo "Englishman in New York" (1987) dir. David Fincher, music by Sting, 1987 (music video) Englishman in New York (2009) dir. Richard Laxton, Leopardo Drama (film).

Levinas, E. 1979. Totality and Infinity an Essay on Exteriority, The Hague: Martinus Nijhoff Publishers.

Said, E.W. 2000. Reflections on Exile and Other Essays, Cambridge, MA: Harvard University Press The Naked Civil Servant, (1975) dir. Jack Gold, BBC (film).

Drawing (...) City (...) Body, Dwelling on Earth – Janeiro (Ed.)
© 2019 Taylor & Francis Group, London, ISBN 978-0-367-02602-8

Dwelling on earth... about the Kampung of Tongkol in Jakarta

M. Louro & V. Catalfamo
CIAUD, Lisbon School of Architecture, Universidade Lisboa, Lisbon, Portugal

ABSTRACT: Dwelling on earth..., impels for a reflection on the sharp population growth and its asymmetrical verification both in terms of time and space. In regards to time, due to the fact that the world population took 200,000 years to reach 1 million inhabitants and in only 200 years reached 7 billion, and is expected to reach 9.7 billion by 2050. In regards to space, this growth will be especially in the most underdeveloped countries, where it will occur in an unregulated and consequently disordered manner. The awareness of this phenomenon founded the reflection established by the research project Houses for a Small Planet, in which the development of the kampung of Tongkol in Jakarta case study was integrated as a fundamental exercise of body consciousness in space, within the building and in the surrounding territory.

With around 30 million people, Jakarta, the capital of Indonesia, is currently one of the largest metropolitan areas in the world. Its population is predicted to increase exponentially over the coming years. This great demographic growth has multiple consequences that are manifested not only in its physical structure but also within its social fabric, exasperating the conditions of basic habitability in the city. The following research on the riverside kampung of Tongkol—an enclave of informality within the megacity of Jakarta—will propose a critical approach seeking to elaborate sustainable and empowering solutions for the emergence of new urbanities.

Keywords: Dwelling, Population, Indonesia, Kampung, Tongkol

1 HOUSES FOR A SMALL PLANET

World population growth in the last decades and its strong concentration in urban areas to detriment of rural areas is a prominent issue nowadays. The research undertaken by Houses for a Small Planet addresses this critical topic and its consequences in urban environments all over the world and it's proposed as a tool to critically rethink about the house of tomorrow, intended as a dignified, man-scaled living space. Hence the aim of the research is to propose sustainable and achievable housing solutions, based on strength and potential of the context and integrated with implementable strategies which could requalify the urban fabric.[1]

1.1 *Research framework*

The various case studies are carried out as academic research embodied in the Master Thesis developed within the framework of the research project Houses for a Small Planet, integrated in the Center for Research in Architecture, Urbanism and Design (CIAUD) of the Faculty of Architecture of the University of Lisbon. The research project was started in 2009 by Professor Margarida Louro and Professor Francisco Oliveira; the first phase focused on the

1. Houses for a Small Planet is a research project started in 2009, with development throughout several phases: 2009–2011, 2012–2014, 2015–2017 and 2018–2020 – underway at the Centre for Research in Architecture, Urbanism and Design from the Faculty of Architecture of the University of Lisbon (CIAUD-FAULisboa). It is a project coordinated by professor Margarida Louro, and involves other researchers from the CIAUD, including professors from FAULisboa, students of the 2nd cycle/master's degree and scholarship holders, as well as cooperation with other national and international institutions.

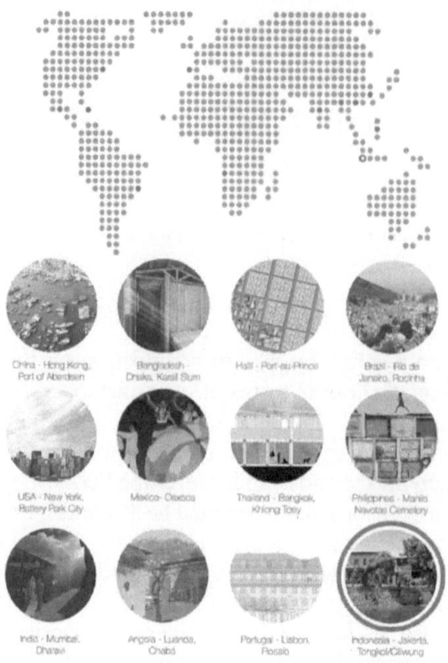

Figure 1. World map with indication of the case studies/synthetic images, Houses for a Small Planet, 2018.

specific context of Luanda and the musseque and resulted in the publication of a book in December 2009, entitled "Casas para um Planeta Pequeno—Projecto Angola Habitar XXI: Modelos Habitacionais em Territórios de Macro Povoamento Informal" (Louro, Margarida; Oliveira, Francisco (coord.); et al., 2009), which involved other professors and researchers from CIAUD. The proposal consists of autonomous and sustainable housing units which introduce a new paradigm of living for Luanda.

In 2010, the investigation project inaugurated a new phase, opening up to students who wish to integrate their Master Thesis into a broader academical research. Until the present time, twelve case studies on different visions and realities have been concluded. Every approach allowed to bring to a common discussion new paradigms, natures and consciences at several scales, being all based on the house—fundamental element of dwelling—as a starting unit in the creation of cities urban fabric.

"Objeto, Edifício, Cidade—Propostas para habitar num planeta pequeno/Object, Building, City—proposals for inhabiting a small planet" (Louro, Margarida (coord.); et al (2017) was recently published and consists of a presentation and a critical confrontation of the eleven project that preceded the case study of riverside kampung of Tongkol in Jakarta (Catalfamo, Vittoria, 2017).[2]

1.2 *Critical reading: The body and the territory*

The different projects in the case studies are tied by a common objective: the proposal of autonomous and sustainable housing units that will enhance local strength, based on imple-

2. From the final master project carried out under the research project: Houses for a small planet. Learning from Informality—Rehabilitation of the kampung of Tongkol along Ciliwung River in Jakarta, developed by the student Vittoria Catalfamo, in the course of the Integrated Master in Architecture specialization in Interiors and Building Rehabilitation of the Faculty of Architecture of the University of Lisbon, under the scientific guidance of the professors Margarida Louro and Francisco Oliveira (evidence defended and unanimously approved in July 2017).

Figure 2. The proposed intervention as seen from the other side of Ciliwung River, Vittoria Catalfamo, 2017.

mentation strategies, and promote solutions of integrated architectural/urban space. Each case exemplifies the compromised living condition in the face of rapid world population growth. This imminent problem is multifaceted, existing in disparate contexts, and therefore the ways it can be dealt with are numerous and may appear sometimes contradictory. However, the act of dwelling is inherent in the very condition of being, remaining unchanged and timeless. It is seen as a common denominator that crosses several physical, political and cultural barriers.

Each strategy is based on a process of thinking and comparing different scales and logic. The proposals for Aberdeen, Bangkok and Dharavi look at the relationship between the scale of the object (boat/container/kiln) and the city. The project for Karail and Chabá focus on the creation of territory, using an analogy of the housing unit as a kind of urban DNA. The studies in Port-au-Prince and Lisbon explores historical memory (mesh, block and limit), in respect to the architectural and urban object. In Rocinha and Oaxaca, the structure of topography was researched, in regards to the strategy between scales of public and private space. Finally, the proposals for Battery Park City and Navotas Cemetery investigate the strategy of layers and the definition of different spatial and living experiences.

The critical synthesis on these eleven case studies provides a wide range of contextual and typological responses, which helped to develop the critical thinking formalized later in the reflection on the riverside kampung of Tongkol.

2 THE KAMPUNG OF TONGKOL IN JAKARTA

Indonesia is the largest archipelagic country in the world with more than 17,000 islands around the equator in Southeast Asia. The capital Jakarta is an extensive urban area on the island of Java, the country's third largest island. The city lies in a bay on the north coast, facing the Java Sea, while green hills and high peaks of the volcanic mountains of Parahyangan flank its south side.

The shift from rural to urban societies is creating a massive impact on the economic, social, political and environmental landscape of countries around the world. Therefore, humanity is facing the challenge of defining the course of this extreme urbanization in order to make it an opportunity for change towards a more sustainable and equitable path than that which exists at present.

The process of urbanization that has occurred over several decades in Europe and North America is happening in only a few years in East Asia, as shown by the emergence of megacities and hundreds of small and medium-sized urban settlements. The region is predicted to continue to be rapidly urbanized as economies shift from an agricultural-based to an industrial and tertiary characterization, with several hundred million people migrating to cities.

The unprecedented growth of a socially and economically divided city like Jakarta offers significant opportunities for a sustainable urban development towards reducing poverty and enhancing resilience. Nevertheless, the city faces a series of difficult challenges in sectors such as transport, housing, water supply and sanitation, and solid waste management. Moreover

Figure 3. Image of the Jakarta Bay with delimitation of the area of Batavia and Tongkol, Vittoria Catalfamo, 2017.

climate change presents an additional threat and the city is increasingly vulnerable to disasters such as flooding and other extreme weather events. As a result, informal settlement dwellers, the most vulnerable class within the city, are subjected to both the hostilities of miserable living conditions and government policies advocating forced evictions, demolitions and relocations.

Although the reality of the kampungs (which in Indonesian means "villages" or "communities") – a denomination currently used to indicate informal territories in Indonesia—is extremely complex and chaotic, it also represents one of the most efficient urban forms of settlement because its informality allows an ad hoc development, according to the physical, social and cultural context in which they are inserted. In this sense, this research explores an antithetical approach to the current practices of exclusion through a visible, flexible, democratic and participatory architectural intervention that acts as a catalyst of urban integration. The final aim of the project is to develop an infrastructural framework that can present not only a tangible improvement in living conditions for the kampung of Tongkol, but also an emblematic attempt to legitimise the place, in turn contributing to the elimination of the constant precariousness that affects the daily lives and future of thousands of people.

2.1 *Description*

The kampung of Tonkgol is located along the riverside of the Ciliwung in the northern district of Jakarta, in the historical area of Kota Tua (in indonesian "old city"), which corresponds to Batavia, the first fortified settlement founded by the Dutch during XVII century. Kota Tua covers two sub-districts of the Municipality of West Jakarta (Tamansari and Tambora) and a sub district in the Municipality of North Jakarta (Penjaringan), with a total area of about 334 hectares.

The Dutch designed Batavia according to their urban planning rules, completing it with a fortress—the Kasteel—and fortified walls, a main public square, churches, tree-lined streets and several blocks separated by canals. This fortified inner city contrasted with the outside where the kampungs, the orchards, and the surrounding rice fields were located; this was the only area in which the native Javanese could live, not included in the planned city due to the fear that they would cause riots and rebellions.

At the end of the eighteenth century the Dutch decided to move south due to the stagnant and unhealthy conditions of the waterways and the epidemic outbreaks of malaria caused by hot and humid climate. They then founded Weltevreden, a more spacious, green and healthy settlement. The wall and the Castle of Batavia were demolished to reuse the materials in the construction of the new institutional and civil buildings, turning Batavia into an abandoned area in ruins.

Over the following centuries there have been several attempts to re-establish the former importance of the city centre but despite recent efforts, most of the policies adopted proved to be ineffective, penalized mainly by issues such as difficulty of access the core of the neighbourhood, pollution associated with traffic and stagnant canals, and the high cost of renovation of individual buildings. The rules implemented to preserve Kota Tua now appear to lack the flexibility to generate the income and interests necessary for its transformation into an economically and culturally vibrant district.

Situated in the middle of the axe that links downtown Jakarta with its waterfront area, Kota Tua has the potential to redeem importance within the city and region. The pressure of development will certainly try to replicate the same model of growth in height with blocks of residential towers as is happening in other parts of the city, allowing for processes of rapid eradication of kampung. Despite the high level of abandonment to which Kota Tua is subject nowadays, it is still possible to find traces of its elegant past and is not hard to imagine the face of the "Queen of the East", as it was called in its golden period, a place like no other in Southeast Asia.

If the Municipality does not act immediately to propose a development plan that preserves the spatial qualities of this heritage, Jakarta, a city with ancient roots, will be another example of self-destruction of its own identity, smashed by new urban realities tediously similar in vision, lacking any construction and spatial qualities and sensibility for the needs of its dwellers.

As previously mentioned, the word kampung has its roots in the rural Javanese tradition meaning "village", but nowadays is often used to denote the neighbourhoods that extend in the back of the official roads and maintains a social structure similar to that of the village; moreover, this word brings with it a sense of cohesion and community, a feeling of belonging. In Jakarta, an estimated 10 million people live in these dense residential areas, behind official roads, infiltrating the formal fabric of the city.

The first indication when entering a kampung is the narrowing of the streets, no longer wide enough to allow access by cars, dwellings squeezed next to each other. This gives the sensation of getting, suddenly, into people's private lives. It is this context that most visitors experience by accidentally strolling through the streets of Tongkol. Along Jalan Tongkol there are two side streets of approximately 4 meters in width, lined with numerous industrial buildings, which act as a filter of transition from the busy urban environment to the cosy atmosphere of the kampung neighbourhood. The last entrance, just as hidden as the previous two, is carried through a large area owned by the Ditpalad Military Command, under a viaduct. Crossing this muddy area, currently occupied almost entirely by large lorries from transport companies, lies the Graanpakhuizen, a warehouse once used to store corn, rice, beans, peanuts, peas and other foodstuffs, which was built by Dutch Company of the East Indies (VOC) between 1700 and 1752 in the east zone between the city of Batavia and the Castle. This historical building was completely abandoned and is now very run down. On the East side of the warehouse, through a small gap between some wooden houses, a narrow, unpaved road leads to the rear of the settlement; this space is flanked on the left by the only remaining part of the fortified Walls of Batavia Castle, demolished by Marshall Daendels in 1809. The Wall, with a height of about 15 meters, has an interesting structure, containing a closed space for guards to hide while safeguarding the perimeter of the fortress.

The streets and direct access to them are the most important aspect of life in kampung. The whole system of social and economic support develops in the streets and, in a way, thanks to them. The street is the central point of socialization, where groups of inhabitants gather at all hours and children play freely. Almost all meals are also prepared and consumed in the streets, permeating the air with delicious sweet smells mixed with some touch of spicy aromas. There is a strong hierarchy of streets that governs the kampung, going from the wider outer public streets, where most residents spend their daily lives working in the informal sector, to the narrower, unpaved private alleyways where vendors constantly circulate up and down selling food. The presence of the River Anak Kali Ciliwung situated along the main alley is still strong in the social construction of the neighbourhood, despite its pollution that dwellers are trying to control. It acts as a natural space with many trees that provide shade and a public space essential in daily community life but also as a potential threat because of the floods.

In physical appearance, the kampung of Tongkol appears as a cohesive, compact neighbourhood with narrow housing units, located side by side, growing close to the ground (usually one or two stories). Some buildings are clearly the results of many layers of construction. In fact residents add extensions to the building not only because of the direct pressure of family growth but also when the family has acquired some source of additional funding. Building materials used for the houses must meet three main criteria: be low cost, light enough to be transported and small enough to go through the narrow access points and streets of kampung. As a result, most homes are built with bricks, sometimes with some concrete pillars that are used for the structure. The floors are made of beams or slabs and the roofs of corrugated sheets, while tiles are often used on facades, walls or decks. In general, the interior spaces are well maintained, clean, painted and decorated with few elements, with the interior walls painted mainly with bright colours: blue (a colour that helps prevent the presence of mosquitoes), green, purple and yellow. For a family that lives in the kampung it is quite common to rent different floors of their home to other family members or other families, while those that face the main street along the river decide to use the ground floor for commercial purposes.

For most residents, one of the strategies to cope with the lack of space in their homes is spend their daily lives outside the living space. As mentioned earlier, in the kampung there are no specific boundaries between private and public spaces, meaning that some residents expand their territories beyond their own home, so that the public street is incorporated into their own housing space.

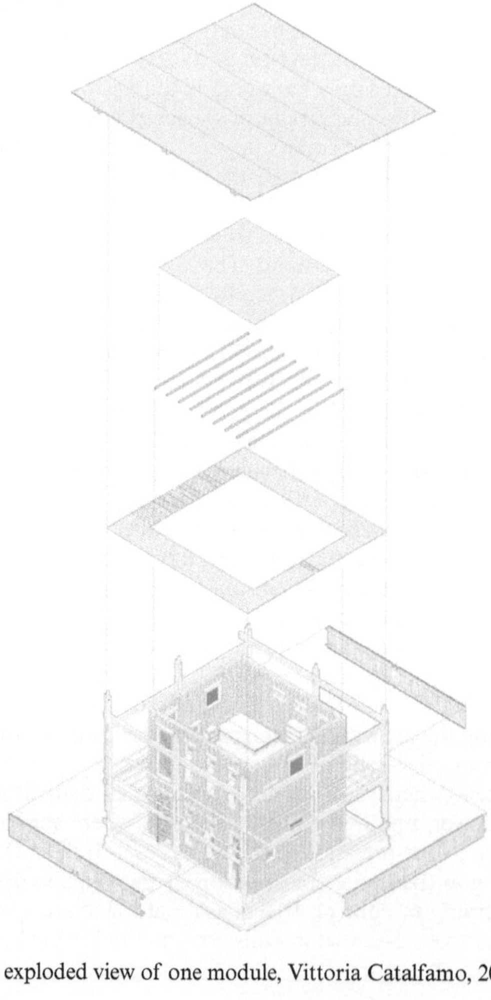

Figure 4. Axonometric exploded view of one module, Vittoria Catalfamo, 2017.

Regardless of the number of inhabitants in the house, people always find ways to handle with the lack of space, using every available space in the most functional manner. Children who grow up, become adults and leave their homes, grandparents pass away, etc ... – the house becomes a flexible concept, ownership does not mean to be a guarantee of stability (in the face of threats of eviction and flooding) and the overcrowding assumes more temporary and dynamic meanings.

Dwellers have developed specific rules and habits to cope daily with restrained space and their uncertain future while at the same time gaining a strong mutual support network and various economic opportunities particular to kampung community lifestyle.

2.2 *Project strategy*

As architects, we literally shape the world and, with it, our lives as well. Assuming this great responsibility we should be conscious that every design choice we undertake brings consequences to people and their way of dwelling. As such, architecture is considered not merely in its formal and physical aspects, but above all as an act of creation and circumscription of a lived space, "consumed" everyday by people, stage and scenography that shapes the everybody's life contributing to the construction of their happiness. At the practical level, this approach translated into a need to live the place in the most direct and intense way possible, hence the trip to Jakarta that represented a unique opportunity, which was consolidated in the choices and consciences of the project. Breathing the life of the community, from the prayer from 5 in the morning to the children bedtime at night, the sense of architecture has become simple, enriched with local wisdom. Every detail, in its form, materiality and location, revealed the reason for its existence.

Due to the impossibility of access to cartographic material, in situ surveys were fundamental. The area of the intervention was analysed using two different approaches: firstly, at neighbourhood scale, the context of Kota Tua was explored through photographic tours, mental maps, sketches, visits to the most significant buildings and walks at various times of that day to understand the different uses and activities; secondly, at the kampung scale, based on the material developed by NGO ASF-Indonesia and the University of Indonesia, and fieldwork in the neighbourhood, through photographs, drawings, sketches, participation in community meetings and conversations with dwellers and members of ASF-Indonesia. Thanks to these local researches and the material developed by the NGO ASF-Indonesia and the University of Indonesia, rigorous bases were built to sustain the project.

> *"As we enter a space, the space enters us, and the experience is essentially an exchange and fusion of the object and the subject. Atmosphere is similarly an exchange between material or existent properties of the place and the immaterial realm of human perception and imagination. Yet, they are not physical 'things' or facts, as they are human experiential 'creations'. Paradoxically, we grasp the atmosphere before we identify its details or understand it intellectually. In fact, we may be completely unable to say anything meaningful about the characteristics of a situation, yet have a firm image, emotive attitude, and recall of it. In the same way, although we do not consciously analyse or understand the interaction of meteorological facts, we grasp the essence of weather at a glance, and it inevitably conditions our mood and intentionality."* (Pallasmaa, 2014: 232).

Throughout the study, there was a constant concern to create a project entirely designed for the needs of the inhabitants, without introducing constructive complexities or formal opulence. Contrary to what might seem, working with the premise of simplicity has made the constitution of the project extremely complex; each choice of design becomes crucial, taking into account, first and foremost, the multiple boundaries imposed by the poverty situation; secondly, the economic and environmental sustainability that the project wants to achieve and, above all, the great challenge and starting point, the dignification of living spaces. Our interest as architects should not be directed away from conventional architecture and cultural issues, but try to conciliate with simplicity the differences between basic needs and aesthetics; ecology and economy; wealth and poverty; utility and poetry.

The proposed housing model resulted organically from an extensive research path, with numerous experiments and compromises, due to the struggle between the will to provide the best possible living conditions and an indispensable need for realism.

The "undercity" of informal settlement of which Tongkol is part of, is seen by the urban majority either as a plague of the urban fabric or completely forgotten. Hence intervening in the kampung has multiple reasons and objectives; one of them is to make residents aware that they already have the means to make a change, while at the same time drawing the city's attention to the conditions in which thousands of people are living. In this context, the intervention is seen as an act of "agitation" of this real and still unheard-of question. Architects act as mediators between the community and the government, providing the former with knowledge and skills which they do not have access to, and the latter with a vision of what could be a realistic and non-capitalist transformation of informal settlements.

The project consists in the creation of three spaces covered by a modular bamboo roof structure: two modules in bricks—housing and hosting—and a community open space lying between them.

The community space plays a vital role in the intervention. Conceptually is born from the idea of a core that support daily living. Functionally, it arises from the real need of a space for community meetings and activities. Spatially, it act a joint space between dwellers and visitors encouraging interaction and mutual learning, with the eventual aim of the growth of the community.

The bamboo porch—result of a broad research about the social context and habits within the kampung—becomes a truly extension of the house, creating covered and shaded liveable spaces. The vertical gardens constitute a green permeable layer that alleviates the modest dimensions of the houses.

Due to its position in the historic Kota Tua, Tongkol has a real potential to become an important touristic spot to be inserted into the tourist routes, a wish that many of the residents expressed for the future. They have already shown to the Municipality their interest in becoming guards and guides of the remains of the Dutch colonial Wall and the VOC warehouse. Currently the kampung is visited by a moderate number of tourists. When it is discovered, the kampung gives equal happiness to the visitors, as to the inhabitants who receive them. The community offers such a welcoming and open atmosphere that tourists end up coming back throughout their stay, attracted by the strong sense of belonging and mutual support found. Thus, in addition to the housing module, the creation of a reception and host space is proposed. This becomes the site to open and re-connect the kampung to the city and thus defend its existence.

Nowadays, the process of designing contemporary domestic space is mainly based on Western spatial patterns. Copy, repetition, standardization have become principles and aspirations on which the world urban environment is constantly based and built. In the society in which we live—from the scale of the object, to that of the house, the neighbourhood, the city and the megacity—the obsessive attention to the exterior and superficial aspect is exasperated, seeking the creation of aesthetically appealing, clean, apparently perfect spaces. As a consequence, the homogenous contemporary city is the mirror of this process of "beautification" where local specificities, tonalities and diversities that make a place unique cannot survive.

In this way the project acts as a catalyst for change, preserving the unique community spirit of kampung, but at the same time generating income for its inhabitants. In its emblematic

Figure 5. Community module cross-section, Vittoria Catalfamo, 2017.

attempt to create a base to legitimize kampung's existence, it also contributes to the eradication of the precariousness that affects the daily life and future of thousands of people.

There was a time where villages protected us from nature. Today nature is our shelter from the city and the harm it could cause. Taking care of the health of our cities is imperative as it is trying to eliminate the distance between the quality of the proposed space and the needs of its users.

For millennia man received benefits from community living, and within the community built man—scaled spaces. In the contemporary city the human scale is in danger of extinction. Therefore it is not enough to produce the city for its inhabitants, but it is fundamental that they get involved in the creation and qualification of spaces in which they wish to live.

REFERENCES

Catalfamo, V. 2017. Casas para um planeta pequeno. Learning from Informality—Rehabilitation of the kampung of Tongkol along Ciliwung River in Jakarta, Projeto para a obtenção do grau de mestre em arquitetura com especialização em Interiores e Reabilitação do Edificado, Lisboa, Faculdade de Arquitetura da Universidade Técnica de Lisboa.

Evawani, E. 2016. Coping with crowding in high-density kampung housing of Jakarta, Archnet-IJAR. 10. 195–212. 10.26687/archnet-ijar.v10i1.790.

Louro, M. (coord.); et al. 2017. Objeto, Edifício, Cidade—propostas para habitar num planeta pequeno, Object, Building, City—proposals for inhabiting a small planet, Lisboa, By the Book.

Louro, M.; Oliveira, F. 2009. Casas para um Planeta Pequeno—Projecto Angola Habitar XXI: Modelos Habitacionais em Territórios de Macro Povoamento Informal, Lisboa: PixelPrint/Pandora.

Pallasmaa, J. 2014. Space, Place and Atmosphere. Emotion and Peripheral perception in architectural experience.

Drawing (…) City (…) Body, Dwelling on Earth – Janeiro (Ed.)

Body in the space: Eduardo Gageiro's "drawing" photography

M. Baptista-Bastos
CIAUD, Lisbon School of Architecture, Universidade Lisboa, Lisbon, Portugal

ABSTRACT: This paper consists of relating photography with the body through a composition and methodology similar to the art of drawing. Looking at the work of Eduardo Gageiro—a Portuguese professional photographer, internationally recognized—these relations are almost direct, and while analysing his work we can select some elements that are referred as his memories, in which the relationship between themselves is inseparable from his life and work. Reports of an interview made by the author of this paper with Eduardo Gageiro will make a direct connection between the type of investigation it's pretended to show and what exactly the photographer wants to see expressed in his work. After the interview it is made an analysis of his work, with some intended objectives, and through his most revelling photographs, it is intended to show that throughout many years Eduardo Gagueiro's work contains elements identical to the artistic composition of a drawing. Although in photography the result is immediate—due to technology, where the image emerges immediately after capture—in a simple drawing this is just not possible. It also reveals to us that the body as narrator is inseparable from his art. The drawing in its origin is always the main communicator of all art forms, even for the compositional process of a photograph.

Keywords: Art, Photography, Body, Image, Memory

1 INTRODUCTION

Watching carefully the photographs of Eduardo Gageiro (1935–), we almost never see images without the presence of people, because "the body" is inseparable from his composition, being an inseparable part of his art throughout his professional career. "But is there art without people? ... I think not! Is there architecture without people? ... No!" (Gageiro, Eduardo: 2017). His artistic expression is unique in the Portuguese cultural context: for the first time photojournalism has passed into the status of Art, comparable to drawing, painting and other artistic disciplines, where the demand for the "work of art" in its fullness is an obligation of any artist for himself and simultaneously as a social duty. Photographs in the newspapers were merely a simple support for written news and Eduardo Gageiro transformed it into an enormous artistic paradigm, an immediate and perfect drawing: a drawn photograph. "It is rumored that they walk with half of their eyes between their hands and do not regard their separate body from the World, like children and animals." (Jorge, 2008: 13). Body and space, drawn as a single element, inseparable from the world's perception. A perfect and harmonious combination, defining qualities and consequently artistic value to the apparent immediacy of photographed journalism.

> "I like people and to photograph them: I observe and react to their behaviour through my work. I work with joy and sadness, trying to convey not only these feelings, but what I interpret in the moment about them, by reducing the maximum to the minimum, without ever excluding anything." (Gageiro, Eduardo, 2017).

Figures 1 and 2. First image, a woman stands out to us, in an enormous effort and suffering, pulling a fishing boat with a rope along with other men behind—purposely, the boat is not seen in the composition, accentuating the drama and tragedy of this photograph (Gageiro, Eduardo, Nazaré, 1962) – and second image, presents a soldier removing the portrait of the dictator Salazar, on the day of the revolution of April 25, 1975, at the Carmo Quarters in Lisbon—the intensity of this photograph is accentuated by the different proportions between the soldier's body and the portrait, in the precarious balance of both. (Gageiro, Eduardo, Lisbon, 1975).

2 HUMANITY AND ITS CONDITION

2.1 *Type the human condition as a phenomenon of artistic development*

The vulnerability of the precarious condition that Man has in the world has always sensitized Eduardo Gageiro since his childhood and consequently, this was reflected in his work. The world conceived in its art, almost always represents the human being in its contingencies. His childhood spent in Sacavém—an old industrial area on the outskirts of Lisbon since the mid-nineteenth century—was marked by the consequent misery that was daily watched, since at that time there were no kinds of perks for factory workers, with miserable conditions. There was a large manufacturing unit in this locality, the famous "Fábrica de Loiça de Sacavém", a famous ceramic production plant with seventy thousand square meters, that deeply marked the everyday life of this population. Nowadays it's the headquarters of the "Museum of Sacavém". This factory has become internationally recognized, with pieces from its productions, in several public and private collections all over the world. "My parents had a small tavern in front of this factory, and I watched the whole misery... the precarious conditions that the workers had and the great gap of social classes between those who had many economic possibilities and those who were in poverty. At that time this difference was remarkable!" Suffering is notorious and deeply expressed throughout the composition of his work: the bodies framed in his photographs often represent the expression of a conformed and contained suffering, as we find in the example of the famous photo taken in Nazaré—a fishing village located north of Lisbon, where the conditions of the fishermen were terrible because the sea is very rough. In this photograph, a woman pulls a boat to land with a rope against the rough sea. This reminds us of Caravaggio's famous painting entitled "The Crucifixion of Saint Peter", where the saint, crucified upside down, is dragged by ropes like the boat from sea to land: here suffering is the main lead that elaborates the photographer's artistic composition, as if it was a painting or a drawing, and so the treatment and the artistic result are the same.

In Eduardo Gageiro's work, there is no hierarchy of values between the different processes and themes in photography, whether it's more elaborate or instantaneous, the state of art remains the same under any circumstance.

> "I think that when we photograph, we have to feel all the energy of the moment that surrounds us and absorb it, in order to transform that moment into an event. Photography helps us to better understand the actions and feelings around us, in an attempt to improve the world. For me it is a duty, without any kind of religious mission, but with a certain professional religiosity." (Gageiro, Eduardo, 2017).

3 CALM INTRANQUILITY

3.1 *Body in photography*

Deeply marked by an ideal of a better world, more egalitarian and with an inseparable aesthetic of this progressive ethic, the photographer recasts the world filtered by bodies framed within the photographic paper, composed and circumscribed to the composition in his work. The clash that the observer has when he looks at his photographs, begins by analysing the context where the main story is exposed, being however unable to leave without seeing all the details contained in the moment captured by the camera. As it happens during the act of composing a drawing, a painting, a sculpture, or a project of architecture, the grand gesture precedes everything: the great moment, or the perfect moment where the bodies navigate in their photos. The composition of the image is an inseparable part of his work. There is no photo that is not perfectly adjusted to the theme, elevating photojournalism in Portugal to the status of Art for the first time.

He began his work, still a child, at the Sacavém Factory. This mark of child labour that made him apprehend the difficulties of the people in his social context and the simultaneously disturbing and difficult emotions felt in the terrible labour environment of the first half of the twentieth century in the metropolitan area of Lisbon, placed him in a position of observer of the world and of things: photography for him is not a window but a channel transmitting a certain melancholic ambiguity, characteristic of the contained suffering of the Portuguese people, so well portrayed in his photographs, crystallizing the main moments of contemporary Portuguese history: "We have to feel what we are doing... to feel with all our senses and some others, which we sometimes discover while we work" (Gageiro, Eduardo, 2017).

> "People are part of my work, because the tragic moments that I have witnessed since childhood have marked me for the rest of my life ... the exhausted bodies of the workers leaving the factory after dozens of work hours and all that misery without any conditions; those almost inert bodies, all of that was very touching. Yet there was an enormous dignity in those silent and resigned people. They looked like living statues." (Gageiro, Eduardo, 2017).

4 INTRANQUILITY PHOTOGRAPHED IMAGE AS BODY INTERPRETATION

4.1 *Antology of the body framed at the precise moment*

Photography, according to Eduardo Gageiro, is decoded by methodologies very similar to those used in a drawing or in a painting, always inserted in an architectural composition. This method makes his work a harmonious anthology of several aesthetics, translated in the photo-reportage. The rapidity of decision that one must have when an event arises, to capture the exact moment in the deepest feeling, condensed into the most precious second of the instant, immediately transforming journalistic photography into art, without the observer noticing the masterpiece. This simplicity hides an intense study of artistic proportions and composition.

In photojournalism, the immediate sometimes becomes art, due to its technology, the quality of photography and the sensitivity of the photojournalist: in the case of Gageiro, it is always art: something rare in his profession, because in other disciplines related to artistic activity, time is fundamental to complete the process, whereas in the universe of the image photographed, immediacy is the temporal totality that the photojournalist has to capture the precise second in his objective lens. The bodies exposed in the photographs of Gageiro, are also unrepeatable, that is to say that each photograph captures a single moment; however, in the same historical and temporal circumstance, hundreds of photographs can be taken, as is the example of his famous images taken in Lisbon on the day of the Revolution of April 25, 1975 – in the same story, each sequence becomes totally different, telling its own story during the event.

This ability to enlarge the moment captured almost until its emotional exhaustion, has a component heavily dependent on the rigor and bitterness retained from its most remote memories.

"When I was a child I spent my time working in the "Fábrica de Loiça de Sacavém", where I witness from an early age the difficulties that many of the workers had afterwards, but I was always aware of what was done in the artistic and painting sections of the factory. A master worker who worked in this department noticed me and began to teach me art and composition in a disciplined way, which struck me deeply; his name was Armando Mesquita, a much respected sculptor in the factory, because he was the one who elaborated the models for the numerous pieces of pottery. I have the highest consideration and respect for him. If he had not supported me at that point in my life, I would probably have gone the other way." (Gageiro, Eduardo, 2017).

From that moment, a melting pot was created in his life, which never returned to be the same. Art and composition unfolded in his way of seeing the body portrayed in the world, with an intensity that is sometimes related to a very intimate and Portuguese spiritual way of seeing the world: the longing transmitted in the gesture, in the movement and in the consequent stop of the bodies, suspended forever in the luminous reflex of the photographic camera, where time stopped.

The exact moment when he takes the photograph appears to be random, but always obeys a rigorous composition, where a dance of the bodies, inserted inside each photo is frozen in the most perfect second: A dance of bodies suspended in time and space, representing what is most important is the spirit of the place, as Norberg-Schulz says "In order to inhabit the earth and the sky, man must understand these two elements and their interaction. The word "understand" does not come here as a scientific knowledge, but an existential concept, which presumes the experience of meaning" (Norberg-Schulz, 1979: 23). In his work with thousands of photographs, the experience of meaning is a work-in-progress. This meaning, to which Norbert-Schultz refers, is recognizable in the atmosphere captured at the moment where the body gives context to an innovative existence: an architectural narrative in the broad sense of the term, with respect to framing and composition.

> "I do not like opaque moments, I like the obvious, but also the obscure, the light and darkness. I expose what is happening in a place, in a circumstance, at that precise moment. I look for the exactness of the exact." (Gageiro, Eduardo, 2017).

5 ARTISTIC NARRATIVE IN PHOTOGRAPHY

5.1 *Multiple intensities in the same image*

Composition is an inseparable part of his work. A proto-architecture, in the sense of conceiving at the exact moment what is the best body position during the action, so that the frame is the most suitable for the decision made at the moment in which something is photographed. Eduardo Gageiro's photography as an art form and expression presupposes common bonds with other arts, such as painting and sculpture, but mainly relates to architecture for three important reasons: a composition determined by action, an action that contextualizes the event and finally the event crystallized forever through the image. "You know, I cannot take pictures without first having a procedure before the shot. The very act of photographing obeys an earlier narrative, beginning with the physical placement of the photographer, who has to be the most appropriate during the observed event, and after that, choosing the right moment for the perfect photograph!" (Gageiro, Eduardo, 2017).

This compositional discovery brings another report to the art of photography: a narrative that enters the patterns that resemble an architectural composition, freed of constraints, without the legalistic authoritarianism that always occurs in architecture, often limiting the professional exercise of the architect. Photography contains a compositional freedom, without legal regulations and ergonomic standards, in which the relation of the body to space is not demarcated by different proportions, because they are both "drawn" with the same resource, in order for the image frozen through the photo to have the importance of the moment captured. Sometimes the bodies are oversized in relation to the context, in others the opposite happens because the composition is dependent on the circumstance of the news that is intended to transmit through the photographed image. The circumstance is what determines the proportions in the news and consequently the photographed image.

Therefore, the image, locked in a photograph, follows the moment and a set of unconscious rules, only the notion in the proportions of the bodies before the framing diverges in different developments from the architecture. However, they contain the similar principle that is the demand of the beautiful in Art.

> "I really like architecture, to photograph the city of Lisbon, which has fantastic buildings and a unique light, but always with the human component: the body is inseparable from architecture and they often appear in my work: body and space, buildings, people, that is, the world seen through a lens, reflecting the intensity that photography allows." (Gageiro, Eduardo, 2017).

6 IMAGE AND CONTEST

6.1 *Censorship and freedom of imagination*

"Imagination is the breaking of a rule hitherto considered inflexible—a rule formerly created by man. Succeeding another new rule" (Lucretius, 2014: 15).

The organization and structure of his work insert a balanced mixture between the exposure of the event and a subliminal message. Nothing is totally exposed and sometimes there is a hidden story in the main theme of each photograph. This methodology was perfected over dozens of years, since part of his professional career was marked by censorship promoted by Salazar's dictatorship, which was a fascist, autocratic and state-corporatist political regime that ruled in Portugal for 41 years without interruption, from the approval of the 1933 Constitution to its overthrow by the Revolution of April 25, 1974. It was one of the longest in the twentieth century. This regime tried to control by force all the social media, artistic and any type of public intervention that could doubt the Status Quo.

Since all of his photos went through an analysis of censorship before being published, Gageiro sought escape strategies to avoid disapproval, finding through subtle forms the power to criticize misery and all the social imbalances that existed, without the censors noticing: the artistic and compositional intensity in his work contains a sublime beauty, which disguised everything else. "I disguised, but I hid nothing! ... because I wanted to show the misery that the country went through in this terrible period of Portugal's history!" (Gageiro, Eduardo: 2017). The existence of a classical meaning in his art through a search for Beauty, mixed with the tragic sense of life, brings about a phenomenological approach in his work, somewhat in the image of Husserl or Merleaux-Ponty, through a transcendence, as Lyotard underlines: "A consciousness conceived of logically, but an actual consciousness." (Lyotard, 51). "The domain of experience as absolute essence, is by essence independent of all being pertaining to the world, to nature, and does not require even these for its existence." (Lyotard. 52).

> "I had to find other ways to show the serious social problems that the country was going through and aesthetics was a way of getting around Fascism: a disgusting organization that controlled people's lives. I composed the photographs, with the

slimmest possible frames, in the image of the great classical paintings. This way, beauty could carry other terrible information about the political situation that the country was going through." (Gageiro, Eduardo, 2017).

7 BODY ABSENCE/LACK OF LIFE

7.1 *Body or not: That is the issue*

"There is no Gageiro's photograph where there is no movement." (Baptista-Bastos in Vasconcelos, António-Pedro and Ferreira, Leandro, 2014).

The body becomes part of the composition throughout his work, except for a very complicated period of his life, when he was diagnosed with cancer. Shocked by the news about his life being reduced to a short term, he began compiling a final book. "During this very difficult period of my life, where I came face-to-face with death, I thought of organizing a last photo book which I called "Silences." A farewell book." (Gageiro, Eduardo: 2017). Throughout the time that he was in his archive selecting the most appropriate photos for the book, he came across with this extraordinary phenomenon: "I like photographs with people, but my mental and physical state was so weak that I chose unconsciously photographs without anyone." (Gageiro, Eduardo: 2017).

Given his state of health and despair, when he chose the photos, they almost never contained people. These were lonely and distressing photos. According to him, photographs without people convey a certain desolation. "I selected again and when I chose the photos, there was one or another person, but the last one was an isolated cemetery in the middle of a field that represented death. I was so sad that I almost excluded the human being from this book." (Gageiro, Eduardo: 2017).

However, after chemotherapy treatments, a unique story takes place in his life. The cancer suddenly regressed and after knowing that he was cured, he changes the end of his book in an extraordinary way: "I had the book almost finished, but upon learning of this news, I was so happy that switch the end and put a picture with a person walking on a road to the horizon as a sign of hope (laughs)! ... We must always have the greatest hope in people! Humanity is the last form of hope in this world." (Gageiro, Eduardo: 2017).

The inseparability of the creative act, life and humanity, are essential for the construction of an imaginary. In Gageiro, the relationship between the human being and his art is unbreakable.

> "Throughout our lives, we have accumulated episodes and probably when I photograph, these episodes all boil down to one, which is the deep respect I have for humanity and the attempt to reflect their dualisms—the grandiosity and pettiness—though photography." (Gageiro, Eduardo, 2017).

Figures 3 and 4. First image portraits the start of construction of the University City Campus in Lisbon—the workers are going home after a long day. (Gageiro, Eduardo. Lisboa, 1969) and second image shows Dictador Salazar at the Sain John's Fort in Estoril. (Gageiro, Eduardo. Lisbon, 1962).

8 CONCLUSION

8.1 *Body instant art work*

His work portraits reflections of memory at a specific time. This reflection is modelled for the photograph, where composition is a hidden element, but fundamental and always present in its professional path. This elaborate arrangement, in the circumstance of the moment, is immediately eternalized in the image.

Photography grounds History, which is apparently explicit to the eye of the beholder. However, the elaborate formal and artistic play between the diversified bodies and the spaces that tell us this story, is the implicit area where Eduardo Gageiro models his art of conceiving an image. For this process, he resorts not only to his memory, but also to artistic canons that are linked to sculpture, painting, architecture, poetry and literature, being the great pioneer in placing journalistic photography at the highest level in Portuguese culture.

His work should be more studied in Portuguese architecture schools, where drawing by hand is fundamental, because it is closely related to environments, where the human body and its context are treated as one element: More than a photograph, it carries the load of an instantaneous representation that always carries a powerful message.

There are no bodies left at random into the space: everything exists at the maximum rigor of a second and the most astonishing is that his work appears a simplicity that hides an infinite complexity. This elaborate spontaneity is picked up quickly with the touch of a finger on a machine that produces instant images. The work of Eduardo Gageiro is above all else the image of photojournalism revealed as a work of art a—a sudden and unexpected elaboration of a crystallized reflection, forever.

> "I try to capture in photography everything I need to tell the same story in multiple ways, but always with a common conclusion: A configuration that is not monotonous, that holds the eye of the spectator, but who is at the same time rigorous, to that we understand the same message in different ways." (Gageiro, Eduardo, 2017).

ACKNOWLEDGEMENT

The author would like to thank Eduardo Gageiro for his good friendship and enormous generosity, through his interview and for allowing to publish his images in this work, and also like to thank Mariana Marmelada for her help in reviewing the translation and formatting this paper.

REFERENCES

Baptista-Bastos, Armando in "Eduardo Gageiro-Um Século Ilustrado". Vasconcelos, António-Pedro and Ferreira, Leandro. Lisboa. 2014.

Baptista-Bastos, M. 2017. Interview with Eduardo Gageiro. Lisboa.

Gageiro, E. 1971. Gente. Eduardo Gageiro. Lisboa: Editorial O Século.

Gageiro, E. 1998. Olhares. Eduardo Gageiro. Lisboa: Author's Edition.

Gageiro, E. 2004. Lisboa no Cais da Memória. Lisboa: Author's Edition.

Gageiro, E. 2006. Fé – Olhares sobre o Sagrado. Lisboa: Author's Edition.

Gageiro, E. 2008. Silêncios. Lisboa: Author's Edition.

Gageiro, E. 2012. Lisboa Amarga e Doce. Lisboa: Author's Edition.

Gageiro, E. 2013. Liberdade. Lisboa: Author's Edition.

Husserl, E. [1902/03] 1999. The Idea of Phenomenology. A translation of Die Idee der Phanomenologie Husserliana II by Lee Hardy, Dordrecht: Kluwer Academic Publishers.

Husserl, E. [1906/07] 2008. Introduction to Logic and Theory of Knowledge. Lectures 1906/07. Translated by Claire Ortiz Hill. Dordrecht: Springer.

Jorge, Lídia in "Silêncios". Eduardo Gageiro. (2008). Lisboa: Author's Edition.

Lucrécio. [50 B.C.] 2014. Da Natureza das coisas. Translated from the original by Luís M.G. Cerqueira. Lisboa: Relógio D'Água.

Lyotard, J-F. 1991. Phenomenology. Albany, NY: State University of New York Press.
Merleau-Ponty, M. [1945] 2005. Phenomenology of Perception. Translated from the French by Colin Smith. Abingdon: Routledge.
Nesbitt, K. 2006. Uma Nova Agenda para a Arquitectura. Antologia Teórica (1965–1995). Translated from the original by Vera Pereira. São Paulo: Cosac Naify.
Norberg-Schulz, C. 1979. Genius Loci. Translated from the original by Anna Maria Norbert-Schulz. Milano: Electa Editrice.
Vasconcelos, A-P and Ferreira, L. 2013. Eduardo Gageiro-Um Século Ilustrado. Lisboa. Produções Panavideo.

Drawing (…) City (…) Body, Dwelling on Earth – Janeiro (Ed.)

Insightful drawings; Invented cities; Inverted bodies

M. Nascimento & F. Khoury
Department of Design and Urban Planning, Faculty of Architecture and Urbanism, University of São Paulo, São Paulo, Brasil

ABSTRACT: This paper presents an articulation between reflections that cover the representation of the body in the architectural space and in the city, "inhabiting the earth" in a transient and unexpected way.

The idea of drawing, as an image of something that holds the memory of an urban event; the idea of the city as a phenomenon constructed from representations that catch passages of "bodies" in its perimeter or horizon, and, finally, the idea of body as an unpredictable element, which acquires meaning and assumes another representation in its diverse enunciation, are issues discussed from the work of artists, critics and philosophers selected as reference for this essay (AGAMBEM, PALLASMAA, VARELA, GALASSI, CALVINO, BRESSON).

The observatory body is dedicated in its detailed contemplation; the traveling body, shifting and projecting itself over what it looks at, in a reversal of roles and complicity between the one who draws and what is drawn; and finally, the artist body, author of ambiguities and imagined landscapes, are the three "bodies" selected by these authors to think the theme of this V Seminar, from a subjective perspective, an intimate connection of the observer and author of images with bodies that astonish perception as insurgent stimuli.

Keywords: body, drawing, city, phenomenology

1 OBSERVER BODY

> "[...] This shop is a dictionary; the language is the system of cheeses as a whole [...] for Mr. Palomar, learning a little of nomenclature is always the first step to take to retain for a moment the things that pass before your eyes. He draws a notepad, a pen, and begins to write names on it and to mark next to each name some qualifier that allows him to recall the image to memory; he even tries to draw a synthetic outline of the form" (Calvino, 1994: 68).
>
> "When Palomar realized how close and error-prone were the criteria of the world in which he believed to find precision and the universal norm, he gradually established a relationship with the world by limiting it to observations of visible forms [...]" (Calvino, 1994: 49).
>
> [...] that is what Monsieur Palomar thinks of himself as a bird. "Only after having known the surface of things," he concludes, "is it possible to search for what is below. But the surface of things is inexhaustible" (Calvino, 1994: 52).

Palomar sees far and multiplies his attention in the numerous layers that scan the selected object in his precise sight. The emblematic figure of Italo Calvino (1994), his name associated with the giant telescope located in an astronomical observatory, the silent and restless gentleman unveils the metaphor of the instrument that gives him function and destiny; and lurks.

The body that observes, with acuity and minutia, immobile in its contemplation of what is permanent, and what is mutant in the landscape, is perhaps the only one responsible for the sensation of being in a specific place, in a time without return. A finite destiny, without recurrence, without future.

Aware of the reactions summoned by the intellect and activated by the sensitivity that comes to us, when in the face of significant stimuli that have altered our relation to events in the world (or environment) in which we are inserted, we can then assume authorship (some will say co authorship) of images, elaborated as thought or sketched with traces and shades in material support, whose peculiarity is that they exist, only from our own observation.

For Varela (1988), the condition of bodily experience defines us and our perceptions and notions, as consequence and origin, limit and quality. According to the author, it is the "cognitive sphere" that houses these perceptive operations and thus defines the conditions of their occurrences:

> "[...] There is no world but the one we experience through the processes that are given to us and that make us who we are. We find ourselves in a cognitive sphere whose borders we cannot go beyond, and from which we can establish neither the beginning nor the modalities ... we cannot bring an experience back to its origins directly. If, in fact, we tried to go back to the source of a perception or an idea, we would find a lot of detail and interdependence. It would always be the perception of a perception of a perception ... or of the description of a description of a description ... There is no point from which we can throw the anchor and say the perception begins here, begins this way" (Varela apud Maldonado, 2014: 40).

Depending on our experiences and experiments, the landscape, as scenery, architecture or place, gains meaning and outline, trace or design, thanks to the selective and critical eye of a peculiar spectator (Fig. 1).

Armed with a telescope, an instrument alluding to the origin of its own name, Palomar assumes itself as providential metalanguage, justifying the attitude of its myopic and astigmatic body in the face of the phenomenon built by the triple opposition between the planets Mars, Saturn and Jupiter.

> "[...] the ancients would believe that they were looking up into the sky of Plato's ideas, or the immaterial space of Euclid's postulates; instead, this image, perhaps by means of which detour, comes to me that we have is too beautiful to be true, too grateful to my imaginary universe to belong to the real world. But perhaps it is

Figure 1. Aquila degli Abruzzi, Italy, 1952. Cartier-Bresson (Source: Henri Cartier-Bresson, 2009: 54). The countless layers of the image; rhythms, murmurs and textures that echo indiscriminately from the distant observer. Bodies that see and are seen, displacements and inversions, the fresh smell of bread and the faded skin of the church wall.

precisely this distrust of our senses that prevents us from feeling at home in the universe. Perhaps the first rule that I must establish is this: stick to what I see." (Calvino, 1994: p. 37–8).

To the contemplative individual, at a literal astronomical distance from his object, there remains only the visual experience. Even when confronted with the image obtained with the naked eye "when he is obliged to take into account the proportions between the planet, the rest of the firmament sparse in the dark space on all sides and he who looks (...)", with the image circumscribed to the lenses of the object that places him as "subject in an illusory face-to-face" with the planet, Palomar's goal is to retain information "on how much of a planet can enter in an eye" (Calvino 1994: 49–40).

Vigilante, susceptible to the doubts that disperse his concentration and constellate his investigatory plot, Palomar, with speculative posture sometimes spontaneous, sometimes programmed; corporeal, or instrument-mediated, is an appropriate metaphor for us to reflect on the inconstancy of experiences and the series of actions and practices of human beings that derive from their perceptions in front of, or meshed in, the architectural building, the city, or the landscape.

These questions, which cross the concerns pointed out by Pallasmaa, when questioning "the dominance of the vision on the other senses, and its consequent predilection in the cognition", also appear in the reflection of David Harvey (1992: 327, apud Pallasmaa, 2011:16), on the postmodern condition: "Aesthetic and cultural practices are peculiarly susceptible to the changing experiences of space and time, precisely because they are involved in the construction of spatial representations and artifacts from the flow of human experience."

Thus, studying and discussing Architecture calls us to consider the body that inhabits, and relates to the architectural expression in varied dimensions and proportions, as a primordial and essential matter, closely linked to the way or circumstance with which this body perceives its inhabitant condition.

From the artifact constructed to the natural panorama or urban event in which the architectural phenomenon manifests itself, any possibility of the conception and use of Architecture must be mediated by the notion of "interdependence between space and time, the dialectic of the external and internal space, of the physical and of the spiritual, material and mental, of unconscious and conscious priorities," "according to Pallasmaa (2001: 17), and thus to establish a strategy whose purpose would be" to rebalance the spheres of the senses, and "to resensual the Architecture through of an enhanced sense of materiality and tactility, texture and weight. Density of space and materialized light". (Pallasmaa, 2011: 36)

2 WANDERING BODY

"[...] I started painting and drawing. And for me, photography was a way of drawing. And that is all. Instant sketch, made with intuition. And you cannot fix it. If you have to correct it, it's in the next image [...]. A photographer can capture the coincidence of lines simply by moving the head a fraction of a millimeter. You can change the perspective with a slight bend on your knees. By placing the camera near or far from the object, the photographer can draw a detail [...]" (Bresson, 1952).

Henri Cartier-Bresson (1908–2004) acquires a Leica camera in 1932 in Marseille. The machine equipped with 35 mm film, whose instant and precise performance transforms the photographic production of the artist, determines not only the nature of the images caught framed in the exact frame, of golden proportions adapted to the original lens, but also testifies the easiness and discretion with which the artist circulates around the city chasing anonymous scenes and faces, distracted by the subtlety of the embedded instrument, Bressonian eye expansion, demanding and restless.

Referring to the new instrument of work as a more sophisticated pencil, whose difference would be to immortalize, in a fraction of seconds, the "drawing" he had in mind, thanks to this new German companion Bresson will practically become a sculptor of image itself, forcing the desired photograph to be submitted to the reality offered by the instrument. Limiting the field of the record obtained to the dimensions of the negative proportions 24X36, the craftsman affirmed that he did not edit his films, impregnated with human expressions and inserted in scenarios and environments he built by his own.

The appreciation of the French photographer for the possibility of constructing the geometry of the image with the new camera refers to the formation received from the teacher and painter André Lhote (1885–1962), from which he received the basic notions of composition and perspective, by frequent visits to the Louvre, getting to know the works of the Renaissance masters, and to the conviviality with purist or pure Cubism artists and the Séction d'Or in Paris in the mid-twentieth century. To the golden ratio was associated the "return to order" of the 1920s, published in the works of the Romanian philosopher and mathematician Matila Ghyka (1881–1965), who published then: "Esthétique de s proportions dans la nature et dans les arts" (1927); "Rites et rythmes pythagoriciens dans le development de la civilisation occidentale" (1931), and "The Geometry of Art and Life" (1946). However, although much of its production is present in the rigor of a geometer, Cartier-Bresson has always recognized and celebrated the surrealist bias of his photographs.

> *"On the other hand, there is no doubt that I see what others do not see. We run, sweat and take pictures. There are photographers like me who have suffered two occupational diseases in the knees and spine. And there are conceptual photographers, who think. This notion of artist is defined by the bourgeoisie of the 19th century. Haydn had to show that his hands were clean as those of the people in the house.*
>
> *They say I'm surreal. No doubt, but few people understand that I am the surrealist of reality." (HCB Interview with Michel Guerrin, published in Le Monde, 8/8/2004).*

We are interested in this article, above all, to reflect on this traveling body, which moves around the city sniffing structures, awaiting possible coincidences and witnessing surprises. A "designer", of characters and places, oblivious to superficial reality, but catching sighs, signs that provide physical and intellectual satisfaction.

Images à la sauvette, the title of his first artist's book, 1952, translated as "muted images", signals the author's irreverence, adopting the idiomatic expression applied to street vendors, unauthorized goods sellers, as the best form to qualify his work, not always in line with the craft as a photojournalist.

> *"[...] But far from dramatizing the turning point, his postwar photographs seek the timeless dimension of the scenes they describe. Like Honoré Daumier's vignettes, with their small groups of characters and simplified scenarios, the photos distill the typical, the ordinary, and the traits of a culture. Like Daumier, Cartier-Bresson was a shrewd observer of social beings, and although he recorded them one by one brilliantly, a huge and extraordinarily rich strip of his work is devoted to groups in public places-from private conversations to hectic crowds. Most are not important people and are doing nothing but being themselves. It is the integrity and autonomy of the Cartier-Bresson photos that throw them from the private to the domain of the most comprehensive metaphor." (Galassi, 2010: 47).*

Friendship with surrealist[1] artists, sometimes resulting in partnerships and cinematic experiences with Jean Renoir and Paul Strand in the 1930s, corroborates to highlight the intuitive and susceptible character of his work, the decisive moment, to the concept of kairos as Jean Clair states, former director of the Musée Picasso in Paris, when referring to retrospective Henri Cartier-Bresson: the Man, the Image the World (2003).

---

1. "He regularly attends the meetings of the Surrealists at the Café Cyrano in the Place Blanche. As a 17-year-old adolescent, he listens to the heated discussions that Breton has with his allies and opponents. Therefore, it are not really the Surrealist painters who leave a mark on Cartier-Bresson, but rather the Surrealist viewpoints on the importance of spontaneous expression and intuition (not luck or chance), and their rebellious lifestyle." (http://www.lecouperet.net/hcb/en/influences).

"Kairos is also used for «a sensitive place» or «a crucial point», including in Homer's Iliad. In this sense, kairos also comprises the art of visual discrimination: almost intuitively—in a fraction of a second—the artist discovers the key point around which he will build his image. [..]. It is exactly this duality—the spatial and the temporal—which makes the term kairos probably the best one to summarize most of Cartier-Bresson's work. It is about the photographer who senses where to make a haunting image and when to fire to give the audience an extra dimension." (http://www.lecouperet.net/hcb/en/influences)

A passing body, moving continually and insinuating by the events that mark the time in the spaces of the city; that define the urban space in the thicknesses of the scenes taken from the silent routine, full of camouflaged visuals in the homogeneous perceptive operations, typical of an attitude accustomed to the daily rhythm.

In a rare filming of the photographer in action, the Albanian filmmaker Gjion Mili (1904–1984) documents Cartier-Bresson's body movement, in an unusual and unpredictable, restless and excited choreography as he watches the New Year Festival in China on the eve of the communist regime's takeover in 1948. Involved by urban courtship, Bresson foresees images, and moves drunken for the pleasure of holding them back in time.

In other circumstances, he orchestrates the projected composition, throw a ball into the air, and capture the immobile lenses of a child's reaction, which tracks the flight of the object excluded from the photographic field. The boy's gaze, suggesting a trance, enlarges the impact of the noisy texture of the worn wall, and reverberates the dissonant harmony between the place and the character, to where all the forces of the movements converge (from the absent object, from the abstracted protagonist and from the conscious author of the desired image).

The work of the French artist is replete with passages and presences caught in the urban plots as accidental events. The silhouette or vestige of an electrifying or suspended body, inhabitant of the invisible mesh of the selected sites to be immortalized by Bresson, is indissoluble signs of the experience and memory that the photographer guards of the city. Signs, or markings, as Agambem (2011) would say; spectra:

"[...] signs, figures or monograms that time scratches on things. A spectrum carries a date, that is, it is an intimately historical being. [...] Thus, in the city, everything that happened in those alleys, in that square, in that street, in those foundations, in that fold, condenses and crystallizes from a blow in a figure, at once labile and demanding, silent and seductive, resentful it is far. That figure is the spectrum or the genius of the place." (2011: 66–7).

The possibility of associating the concept of Agambem with the "disappeared" bodies which, in the next instant, do not exist in the place they occupy in the finished photography refers us to Susan Sontag, and her maxim, when she affirmed that photography is responsible for transforming reality into tautology.

Figure 2. Aquila *Valência, 1933.* Cartier-Bresson (Source: Henri Cartier-Bresson, 2009: 88). The restless body of the child reacting to stimuli from the beyond-frame.

Figure 3. Nápoles, Itália, 1963. Cartier-Bresson (Source: Henri Cartier-Bresson, 2009: 45). An anonymous inhabitant of a city unveiled. Perpendicular tracing, equidistant from the caryatid, embedded in the baluster in dubious testimony: the body of architecture reacts immobile to the transitory body, which balances it in the landscape, unaware of this coexistence.

Figure 4. Hyères, France, 1932. Cartier-Bresson (Source: Henri Cartier-Bresson, 2009: 13). A projectile on two wheels transgresses the movement of the stair spiral, interrupted by the photographic gesture, to last, until, and if, it happens again.

> *"He is a Paris pedestrian himself; dodging through the crowds, always in a hurry, always eager, linking action and word in a single reflex [...] He salutes those who are mourned at Charonne, forgotten at Belleville, exploited at Javel. He is with the poets who are murdered, buried, unknown. He is with all those—on streets, squares, embankments—who form the unspoken, living ideograms of a city whose name is written in the human beings who come to it, whether to make their own names or forget them." (Feyder, 1984, 11–12).*

3 ARTISAN BODY

Palomar observes the photos of Cartier-Bresson extracting countless signs of his multidimensional analysis. Bresson organizes through his lenses the scene waiting for the character who gives meaning to it.

Palomar changes the instruments used to better serve his contemplative intention; Bresson moves in the space to find the exact time of the essential shot.

Calvino (2000), in "Do opaque" (first published in Adelphiana, Milan, 1971) confesses his position as a spectator of the world, assuming that "the properties of space vary according to the directions in which I regard my orientation." (Calvino, 2000: 107)

The place where the body is; its situation before what is revealed and what it supposes to be hidden.

Figure 5. México, 1964, Cartier-Bresson. (Source: Henri Cartier-Bresson, 2009: 52). Intentional ambiguities: the girl conveying the portrait of the elderly woman; the effigy of the symmetrical ancestor, supported by the discontinuous body of the child; diagonal in displacement between the frame superimposed on the knee joint, and the young profile emerging behind the frame; the sophisticated carving around the woman's portrait and the simple landscape around the girl's photograph, about to immerse in the opening of the interrupted fence.

> *"[...]Being the world I am describing a kind of concave amphitheater facing south, and not including the convex face of the amphitheater, supposedly facing north, there is therefore the extreme rarity of the opaque and the widest extent of the sunny" (Calvino, 2000: 115).*

The one who projects, must project himself. Must imagine how the mutual complicity, complementarity and interdependence between drawing, architecture and body occur.

Pallasmaa denounces the absence of plastic and spatial experience in contemporary buildings and architectural environments, deriving from what is known about human existence. (Pallasmaa, 2011; 29)

> *"[...] architecture has adopted the psychological strategy of advertising and instant persuasion; the buildings became visual products disconnected from existential depth and sincerity. David Harvey relates "the loss of temporality and the desire for instantaneous impact" in contemporary expression to the loss of existential depth. FredericJameson uses the notion of "planned superficiality" to describe the contemporary cultural condition and "its fixation on appearances, surfaces and instantaneous impacts that have no retention power over time"." (Pallasmaa, 2011, 29).*

Contemporary philosophers such as Maldonado (2014), who has also been trained as a physician and psychiatrist, has warned us of the possibility of finding in aesthetic experiences fundamentals that help us deal with sensory data through which our brain acquires knowledge about the environment.

Thus, analogous procedures to those of artistic representation could constitute:

> *"[...] the real extension of the ordinary activities of the visual cortex, which are to represent the constant, lasting, essential and stable characteristics of objects, surfaces, faces, situations, and so on, that is, all those operations that enable us to acquire knowledge." (Maldonado, 2014: 108).*

The drawing, and other strategies and means used by the artisan to develop expressive language itself, would therefore be possible resources to recapture, "in a more or less conscious way, the work of selection by which the brain reaches the essential in sensory data" (Maldonado, 2014: 108).

It is also the Italian philosopher who warns us about the symmetry between "the need for the brain to isolate and evaluate the permanent, essential and constant qualities of the objects

Figure 6. Autumn meeting.
Photo: Sandro Miano. Performance "MAR": https://www.youtube.com/watch?v=wLQjCmz5Dss.

of the world and the continuous effort of artistic research to learn to investigate the very essence of reality." (Maldonado, 2014: 108).

A double-hand movement, the one that draws, sometimes invents the body and the landscape that shelters it; and also creates artifice to convince us that it is the clever and skillful body that camouflages and pretends to belong to the space where he has always lived.

The work of the group Urubu Studio, especially the choreography "MAR" by Beth Bastos, amplifies these reflections, as a desired chance to reveal the artifice body. Photographer Sandro Miano, with the participation of other artists, based in a fishing village in Picinguaba, north coast of São Paulo, since 2009, proposes essays and choreographies that explore natural landscapes and environments built to develop the individual's bodily consciousness in the sensitive relationship between time and space.

INTERVIEWS

https://jornalggn.com.br/noticia/a-relacao-de-henri-cartier-bresson-com-a-arte
http://www.lecouperet.net/hcb/en/influences
http://www.oxfordreference.com/view/10.1093/oi/authority.20110803100451553 Sites
https://urubustudio.wordpress.com/2010/05/10/11/
https://run.unl.pt/bitstream/10362/10178/1/Vaz_2013.pdf
http://www.agendadedanca.com.br/encontro-de-outono-com-beth-bastos

REFERENCES

Agambem, G. 2011. Da utilidade e dos inconvenientes de viver entre espectros. In Revista Serrote n° 6, p. 65–9.

Bresson, H. 1952. The Decisive Moment, New York. Copyright 1952 Cartier-Bresson, Verve and Simon and Schuster.

Bresson, H.C. 2009. Henri Cartier-Bresson: fotógrafo. São Paulo: Cosac Naify.

Bresson, H.C. 2010. Henri Cartier-Bresson: o século moderno. Tradução Cid Knipel. São Paulo: Cosac Naify.

Calvino, I. 1994.Palomar. Tradução Ivo Barroso. São Paulo: Companhia das letras.

Calvino, I. 2000. Do opaco.In: O Caminho de San Giovanni. Tradução Roberta Barni. p. 103–119. São Paulo: Companhia das letras.

Clair, J. 1999. Europeans.London: Thames and Hudson.

Feyder, V.; Mandiargues, A.P. 1984. Henri Cartier-Bresson à propôs de Paris. Paris: Thames and Hudson.

Galassi, 2010. Peter. Henri Cartier-Bresson: o século moderno. Trad. Cid Knipel. São Paulo: Cosac Naify.

Maldonado, M. 2014. Da mesma matéria que os sonhos (sobre consciência, racionalidade e livre-arbítrio). São Paulo: Edições SESC São Paulo.

Pallasmaa, J. 2011. Os olhos da Pele (A arquitetura e os sentidos). Tradução Alexandre Salvaterra. Porto Alegre: Bookman.

Sontag, S. 1977. Sobre Fotografia. São Paulo: Companhia das letras.

Varela, F. 1988. "Il circolo creativo: abozzo di uma storia naturale della circolarità". In Watzawick, Paul (org.). La realtà inventata. Contributi al estruttivismo. Milano: Feltrinelli.

Drawing (...) City (...) Body, Dwelling on Earth – Janeiro (Ed.)
© 2019 Taylor & Francis Group, London, ISBN 978-0-367-02602-8

A contra hegemonic design for the Brazilian cities, the Rio de Janeiro case

P. da Luz Moreira
Departamento de Arquitetura, Escola de Arquitetura e Urbanismo, Universidade Federal Fluminense, Rio de Janeiro, Brasil

ABSTRACT: In this 5th Seminar of architectures imagined: graphical representation and other architectural images in São Paulo have space for a reflection on the practices and performances policies with which the Brazil has been confronted over the last few years for conformation of its territory and its cities. Our urban policy. Responsibility of the Ministry of Cities, created in 2002 in Brasilia, and that meant not just repeat the inertial form of reproduction of the Brazilian city, but generate another agglomeration, with more justice and inclusive. The question remains, a review of what we have acquired or we are striving to achieve, an contra hegemonic city plan for Brazil's cities can be imagined, designed and dreamed?

Keywords: hegemony, design, density, inclusion, mobility

1 MACRO POLITICS

Within the more general context, in Brazil and in the world sees the decline of the narrative of neo-liberalism, as discourse of legitimation of the macro policy, from the crisis of the international financial system in 2008. However, it has not yet emerged alternative proposals, which occupy the space left by the doctrine of fiscal austerity, which still seems to govern the actions of the most diverse political tendenciest can be observed, with a return to periods of Felipe Gonzalez (1982–1996) in Spain, Tony Blair (1997–2007) in England, or Françoise Hollande (2012–2017) in France, or even Lula (2002–2010) and Dilma (2011–2016) in Brazil. All ended up doing resonance box of issue of austerity in public accounts, moderating speech, not only in Brazil, but throughout the world. Indeed, that denounces the clear absence of identity of left-wing governments, in recent years, not only in Brazil, but in many parts of the world. This statement has of greater income distribution and by choosing processes, which clearly encouraged the employers to the detriment of the world of work. In the specific field of planning and design, these governments invariably had favored real estate market, waiving of the adjustment of the value of the land by the State, opening up a public management (see Rolnik 2015).

All these forms of government have embarked on a quest for consensus on top, that is. without the participation of the masses impoverished, investing heavily in the model of European social democracy of the second after the war. This is, there was any investment in the radicalization of democracy, in the intense participation of the masses. And, in our field of architects and urbanists, the definition of housing policy and conformation of the territory of the city, space where the offerings of opportunities for the most vulnerable social people appear very clear and even didactic.

> "Those who seek consensus is authoritarian regime, democracy is not. Democracy is the recognition of the conflict, the pursuit of negotiation and seeking agreement, always provisional, in function of the correlation of forces." Fernando Henrique Cardoso, quoted in COUTINHO 1979 page 47.

The plan and the design, while pre-territorial disputes are powerful tools for explanation of these conflicts inherent to democracy, and their radicalization, which could mean an important impetus in the explanation of the diversity of opportunities that exist in our society. And here is of course the great frustration, who represented the coming to power of the Partido dos Trabalhadores (Workers Party PT), which had as its main ideological discourse the fight in the form of operating our oligarchy, the Prussian via[1], after all the party repeated that he wanted to self-government of the workers and the masses excluded by history. The figure of Lula involved a great symbolism, after all it was the very incarnation of this accession to the precariousness of Brazil, union leader who in his appearance was declared against the union contribution required, was the son of immigrants from the northeast, and survived in São Paulo as metallurgical employee of conservative modernization of civil-military dictatorship. But the symbolism of the conquest soon showed itself as continuity in the traditional way to operate the Brazilian elite—the Prussian via—who, despite the promotion of an unusual inclusion, refused to overcome the old exclusionary operation, lacking social magnification, both political and economic. After all, the thesis that "the change will be promoted by the State power, and not by the self-organization of social life", once more won, and also demonstrated to failure. Once more, the governments of the PT were hostages of tariffs perspective of major national banks and the demands of large contractors.

The question is blatantly more clearly, when we are discussing on how to operate the Brazilian State in the configuration of our urban infrastructure and development, which in fact remained privatized and operated for the benefit of a minority, the large contractors. The works of the World Cup of 2014, or investments in urban mobility, or new installations of the oil industry, or even the works of the 2016 Olympics in Rio de Janeiro, finally the structural planning of our cities and our territory are being determined by private interests that even make consultations with large majorities. In the design of the elite leaders of the country the constant neglect by the actions of plan and design, and the predominance of the interests only of implementing agents repeatedly denounced the privatization of the Brazilian state bureaucracy. This form of predatory operate, and which operates on the logic of appropriation of the largest number of immediate benefits, as if the nation had no future or new generations ended up generating and encouraging major corruption scandals, repeating the Prussian path. The present work on the assumption that the actions of socially transparent planning and design are fundamental to the radicalization of democracy, provided that their pre figurations are submitted to a wider universe, conforming a citizen participation.

The decline in the form of act of social democracy in Brazil and in the world, embracing the social regulation of employers, the real estate market, penalizing the world of work and being involved in processes of urban gentrification was a gradual process of long term. He began with the oil crisis of 1973, and its impacts conflict in Israel and the Arab world, with consequences also for the balance of the colonial system led to the first global crisis of the post-war period, disarranging growth and income distribution in that period. Some authors mention a peculiar reversal that the capitalist system was able to operate, with respect to the guideline of May 68, in which there was a clear criticism of the bourgeois state, bureaucratic and disciplinary action. May 1968 refused the status and the institutions of bureaucratic control of both liberal democracy, much of Soviet socialism, revolted against the material reproduction, the devices of social control, from a critical encompassing the capitalism as the economic system and mode of existence, replacing the issue of democracy and of the revolution. However, at the end of the 1970s and early 1980s, the right captained by the Anglo-Saxon world, with Prime Minister Margaret Thatcher of England, and the American President Ronald Reagan take the lead and assume proposals for deregulation and retraction of the State, taking account of partial way the guideline of May 68.

> "The heralds of the current economic model like to see as winners of a clash in which would have demonstrated to the world that the neoliberal capitalism was the

1. COUTINHO 1979 page 42. Points out a way to operate autocratic and undemocratic of our oligarchy as the *Prussian via* an analogy with the processes of unification of Germany.

best way, even the only, to produce wealth, innovation and well-being." SAFATLE 2017 page 23.

In fact it is necessary to recognize that the neoliberal discourse has won many hearts and minds in the world, colonizing not only ways of structuring governments and institutions, but also the everyday life of various agents and actors. Effectively, the increase of individual autonomy, flexibility of ideas daily, criticism of the bureaucratic state, celebration of entrepreneurship were present in agenda of May 68.

> "Currently, we know studies that defend the thesis that the rise of neoliberalism at the end of the year 1970 is a special unfolding of the impulses May 1968." SAFATLE 2017 page 27. "Should assume that the extreme right has been able to establish a political response, listening to the social discontent, the insecurity produced by an economic system of impoverishment and increased vulnerability." SAFATLE 2017 page 32.

From one moment to another, the system began to celebrate the grammar of entrepreneurship, the decline of patterns of social solidarity, the imposition of a competitive individualism which edge the animality, the end of employment, and the relentless pursuit by deregulation. However, with the overwhelming crisis of the capitalist system of 2008 this discourse loses power of persuasion, popping the first surveys, including liberal economists, that poverty is increasing continuously, from the deregulation of the 1980s. In the period 1910–20, the income of the richest 10% accounted for between 45% and 50% of the American national income. This percentage drops to 35% in 1950, reaching 33% in 1970, reversing this trend in the 1990s, and returns to the levels of 1910–20 (45% to 50%) between 2000 and 2010. The author also highlights speeches that laid in check, the moral argument of meritocracy, as the work of two Italians economists; Guglielmo Barone And Sauro Mocetti, which showed how the surnames of rich people in Florence were the same for almost 500 years, from 1427 until 2011. SAFATLE 2017 also underlines the cynicism of employers in these years, in the speech of the billionaire Warren Buffet;

> "Who said that there is no class struggle? It is clear that there is one, and we are winning." Warren Buffet, cited by SAFATLE 2017 page 24.

Even in the field of macro policy seems to crucial for Brazil structuring policies that stimulate a greater balance in the distribution of income in the country, reversing one of the worst concentrations of the world. The structuring of the territory is one of the most powerful tools to distribute opportunities among the citizens of a country, the question of the location is fundamental to guarantee access to education, health, culture and leisure. Activities, which in the space of a generation ensures that families located in the zone of insecurity can overcome their limitations. Therefore, urban policy is a fundamental factor.

A general balance sheet, in the field of urban policy, we came to the conclusion of a huge mediocrity, in our discussions and in our concrete experiences, which do not have outgrown old paradigms are installed, which are repeated, without at least being offered an alternative. The Brazilian cities, space where inhabit 85% of the national population, approximately 170 million people remain trapped in a housing production, which reproduces the old BNH[2] practices, such as the Minha Casa, Minha Vida[3] (MC, MV). Or in the area of urban mobility, where we reproduce so little creative programs of systems for Bus Rapid Transit (BRTs) in an uncoordinated manner, without obtaining a structured action between various transformations, that the urban always requires. In an election year of 2018 the urban question, which today afflicts millions of Brazilians, approximately 170 million, is not guided and there is no forecast to be discussed in the presidential campaign of a clear and articulated. The main candidates do not debate the space built in Brazil, and the alternatives to its current form to

2. BNH acronim in portuguese that means National Housing Bank.
3. Minha Casa, Minha Vida, literaly My House, My Life it is a government program of housing in Partido dos Trabalhadores (Work Party) years.

reproduce and expand. The Brazilian city is still produced generating; ghettos of poor and rich ghettos that protect themselves, avoiding any mutual contamination, a city with their mobility dependent on private cars, a city scattered and dispersed, where the infrastructures are not disseminated, and finally a city that interacts not well with the natural environment. Only these four points not identified and does not formulate any proposal to oppose the inertia of the current production of Brazilian city, this is followed with the Ministry of Cities hostage to a policy that repeats practices of technocratic times of military dictatorship. The housing deficit, which seconds recent studies reaches 6 million houses follows unachievable, with governments producing enterprises distant and devoid of any urbanity, in outlying areas. On the other hand, the former Brazilian urban centers, which invariably focus constructions and remarkable efforts are emptied and abandoned, with its heritage being damaged. The works of mobility are timid and do not penalize the individual automobiles that are receiving subsidies from governments, whether in tax exemption or roadworks. Finally, we urgently need mutations and change our urban policy, staring in front of inertial our way of making cities.

2 IMAGINATION, DESIGN, PLAN AND PROJECT

The imagination, the design and the project constitute themselves as a form of real's approach, which is not restricted to a mere contextualization and quantification of problems, they are investing heavily in the grow or become another, in the constitution of scenarios that are not restricted to reproduction of the current status quo. In its capacity to build scenarios and visions of the future, imagination, the design, the plan and the project depart from approaching the diagnosis and prognosis. Explains the future, more than the analysis of the existing context. Here, there emerges a risk of manipulation of images empty, which is not confronted with the real problems posed by society as a whole, what are your effective claims processing. Hence the need to build a program of needs and expectations, able to explain the general principles which inform the imagination, the design, the plan and the project from which we are (diagnosis), to propose that we want to come-to-be (prognosis). The architecture and the city are as common collective heritage enjoyed by everyone, to comply and represent the collective. When there is participation in the construction of the future scenario we are aware that the management of your routine is as important as its implementation. There is a concept in imagination, in the design, the plan and the project, which the renaissance called adequacy, an ability to fit the proportions of the whole and parts of the proposed scenario with the existing context. A certain balance between ethics and aesthetics.

The intention here is to restore the character of architecture and urbanism as art, understood as a force present and synthetic that live together with their premises; functional, ideological and constructive. In this sense, the word architecture is enlightening when dissected, and its significance connected to an duality with enriching and potentiality

> "So, preceding the expiry tektonicos (Carpenter, manufacturer, building, construction) added to the radical arche (origin, beginning principle)...The arche is the center of social sphere that world, and should be translated in buildings, presenting the gods, the history and the ethical spirit of the Greek people." BRANDÃO 1991 página 22.

The concept of arche balanced principle of the universe, a point of balance between the man and the Kosmos, as a sign synthesis of the ordering of the world by man is the key that opens for us the understanding of the various sensibilities, which will build the idea of modern man. The Arche is a concept that is beyond the materiality of the building, but it is only possible to be unveiled by its own materiality. As a world that the human transformation of nature becomes visible when it is performed with aesthetic concern, therefore distinct from simple construction. Is this concept a explainer triad; first one return to origin, according to an authorizing unit and finally, an expressiveness that gives visibility to the world that it is bound.

3 GENERAL PRINCIPLES

3.1 *Inclusion and celebration of diversity*

Inclusion is the greatest challenge of Brazilian cities, which have a history of continuous passive exclusion of people and areas, which remain as ghettos of poverty, not assisted the most basic infrastructures. After all, one of the most striking characteristics of our society is the remarkable concentration of income. In a general way, our politicians and our policies are not yet awakened to the fact that the territorial distribution of the population may be a factor capable of distributing opportunities, and therefore income. The simple access to a centrality more strongly constituted, can mean the frequency in cultural and educational equipment or good performance, substantially changing the perspective of vulnerable populations. The simple deployment of basic sanitation in certain localities differs significantly the occurrence of diseases such as diphtheria and disarray, that may in the first years of life mean definitive impairments in cognitive ability of individuals.

Diversity is a didactic, attests to this fact the strategy adopted by North American universities, who years ago make a systematic effort to gather in the same classroom students from different backgrounds and nationalities, in the expectation that their shared experiences to form a critical mass. The excellence of the North American university has one of its pillars in this pre determination, which enables an experience of sharing experiences, which ends up producing a learning, where the passivity gives place to activism. The experience of the North American nation[4], who based their development in the attraction of different nationalities, and during the passage of the 20th century represented a promise to the immigration of all peoples. In a way, the New World of America in its entirety also represented this place so emblematic, a place where the opportunities were open to people from the east and the west. The urban operations need to embrace this goal, encouraging the exchange between diversities.

The pedagogy of Paulo Freire, also points in the same direction, diversity is didactic, capable of making us relativize our values, and therefore produces a didactic impulse of relativization of our values. The dialogic theory of Freire 1970 points to the basic premise of the dialog between experiences of any origin as operation didactics, contrasted the authoritarian concept[5] of education, which did not generate autonomy of thinking, but domination and colonization. Here there is a levelling important between the cultures of the colonizer and the colonized, center and periphery, a new proposition of relationship between teacher, student and society. Bring these values to the ordination of physical space from cities, neighborhoods and immediate vicinity is to restore the original direction of human settlements, where diversity have a quality of being educational.

3.2 *Identity and expansion of self-esteem (Portuguese matrix of our space)*

On the other hand, it is fundamental joint in new manner the idea of being and place, as a construct that brings awareness, self-esteem, neighborhood, and community mobilization. In this matter, the concept of analogy is fundamental to structure the demands of users and residents, who often are stuck to a routine so devastating, who cannot overcome their very limited horizon of possibilities. ROSSI 1979 used the concept of city analogous to define the importance that the collective memory of the citizens can play in the process of electing the new design of the city. Inspired by the paintings by Canaletto, the methodology of the city analogy was a surrealist base compositional operation, which from some demands tried to offer a new reality. The analogy always demand a match, or an identity that a human collective shares from a same common cultural basis. Here, this foundation is our Portuguese matrix of city, and more specifically in the case of Rio de Janeiro its interconnection analogous to

4. TOTA (2009) Page 122. Between 1901 and 1920, the United States has received approximately 15 million immigrants.

5. In theory of FREIRE 1970, the name is "concepção bancária" in Portuguese. In the sense of authority.

Lisbon. Always reaffirming the diversity present in Lisbon and in Rio de Janeiro, as plural metropolis, which are not restricted to a single face, but to a wide range of deployments, and therefore personalities.

Important to emphasize that this is not an operation that refers to models conservationists and protectionists, because in spite of the recognition of the single array—Portuguese—the reference points not only for the restoration, but has evolutionary claims and adaptable, which are demanded by contemporary life. Rossi (1977 and 1981) and also Tafuri (1981) always have manipulated the history is not in its descriptive meaning and diagnosis, but in its potential propositional, formulator of a prognosis, able to make potent the project, so its will become.

4 GUIDING PRINCIPLES

4.1 *Dense and compact city, that revaluation his patrimony. Zero tolerance with the expansion of urban stain*

This issue is a huge challenge, to prevent the reproduction of the Brazilian city, fighting their endless dispersion involves the change of mentalities and idealizations deeply rooted in society. The well live socially shared and installed involves an isolated dwelling unifamiliar, next to a natural idyll, without proximity to neighbors. This condition was achieved from the hegemony of the North American city in the world, who went on to represent for the common sense the standard of living well. This American cultural hegemony, which emerges from the end of the first world war is a complex phenomenon, which was examined by the Italian philosopher Antonio Gramsci, in the Prison Notebooks, set of manuscripts produced in the fascist prison, which was submitted.

In this context, the rise of the North American hegemony in the world, with the strengthening of individualism, fighting class associations, the Puritan adjustment of sexual habits and alcool behavior the laborers, and also the high salaries and the granting of social benefits, which characterized the fordism was also announced. In the book 22, the philosopher of Sardinia discusses the Americanism and Fordism, to understand the emergence of what ranked as "the most advanced capitalist social formation" of his contemporaneity. The end of the first world war marks the rise of the domain of the United States in the world, this fact will be reflected political, cultural and economic rights for the history of the 20th century. Gramsci is a critic of everyday mores conformed by dominant ideologies, investigating the establishment of common sense in the consolidation of habits and gestures naturalized. The American way of life ended up winning mentality in various parts of the world, including Europe, changing our shared conception of what we consider the living well. Gramsci had a critical view on the americanism and Fordism, but also showed a transforming capability;

> "What is now called americanism is in large part the critical maintenance of old strata that will be discarded by the possible new order, and who are already taken by a wave of panic, social dissolution, of despair, which is an attempt of unconscious reaction of who is powerless to rebuild and leverage the negative aspects of the transformation. There is social groups condemned by the new order which can be expected to reconstruction, but those who are creating, by imposition and with his own suffering, the material bases of this new order. These must find the original system of life and not American brand, to make freedom what today is need." GRAMSCI 2008 page 89.

In the field of architecture and urbanism these consequences are materialized in human form to occupy the territory of our planet, which lost strength from the oil crisis of the 1970s, but still apply so active in different mentalities, and continue to be naturalised in various parts of the world. The road celebration, determined and still determines the spaciousness of our cities, with the expansion of areas dedicated to cars, the deployment of expressways, viaducts and works that have increased the reception to the displacement on tires, penalizing public life and circular of pedestrians. In addition, the american city disseminated the paradigm of dwelling unifamiliar of low density, near idílios isolationist, the suburbs mono functional amenities, surrounded by nature. The two determinations, the hegemony of road celebration and dwelling unifamiliar isolation on the fringes of the city, just representing a huge fraying of the urban fabric, lowering the density and the proximity between citizens. These representations of living well continue impose the mentality of expressive of contemporary population quotas determining strong environmental impacts, and excessive costs for the universalization of urban infrastructure.

Therefore, the inertia of the real estate market, and the generalized behavior of our society, that when thinks in live, imagine a secluded and idyllic retreat close to nature, with a garage full of cars. This attitude determines the abandonment of huge urban voids with infrastructure already, that need to be reocupy, for which we will start facing the problem of the universalization of urban services such as water supply, sewage, garbage collection, lighting, transport networks, rainwater, etc... Anyway, everything that constitutes the urban amenities that we enjoyed in the areas with most infrastructures, and that demand public investments invariably or managed by public spirit. When the stain is dispersed and scattered in the territory more difficult will be the presence of all these facilities in all parts of its territory, therefore the dense and compact city is much more sustainable from an environmental point of view.

The growth rate of urban stain of the city of Rio de Janeiro, pointed out by the architect and urbanist Vincent Loureiro, coordinator of the Metropolitan Chamber[6] is approximately 30 square kilometers per year, which demonstrates empirically the preference, both from the market, much of the population by the conquest of new areas, unlike the joint ventures on old quarters. Therefore, the fight against rampant expansion of urban stain is difficult and must be addressed in the context of changing conceptions and mentalities individuals, who represents the living well. While the living well, for the common sense, is represented by the single family housing unit built in the vicinity of nature, dependent upon a disabled centered on private cars, we will continue with the criminal expansion of our cities. Such an attitude, in addition to impacting strongly the environment, also condemns significant portions of the population to live without services that characterize the urbanity.

The issue is complex, and is charging of civil society, the entrepreneurs and the public a state policy that is able to contain the seduction of new real estate ventures coming to continuous natural. Zero tolerance with the expansion of urban stain. The proposition alternative should try to coopt and attract economic activities capable of offering a tempting option, for example, encouraging the deployment of farms and productive sites, based on family farming, producing food for the supply of the city of Rio de Janeiro. In this sense, the Rio de Janeiro suffers from major shortcomings, recent surveys tell us that less than 30% of the foods consumed in the picnic schoolchildren of the metropolitan city are from the fringes and limits of urban spot. What shows us that there is a profitable economic activity capable of facing the predatory real estate speculation, and tuned with a little more friendly to the environment.

In addition to this activity on the fringes and limits it is necessary to revitalize and care for it properly to our heritage built, which has copies expressive in different centralities of large agglomeration of the city of Rio de Janeiro. The transformation of uses of old structures must be fostered, attracting preferably using housing, in order to constitute multi functional centralities, who take advantage of the infrastructure already present in these contexts, ensuring citizen movement and life, even at the end of the week. The housing density should be directed to the existing standards in specific cities, such as Paris, San Francisco, Lisbon, New York, among others. Besides the issue of multi functional, must be guaranteed access to the Centralities extracts more vulnerable economically, always seeking a neighborhood with a variety of social classes.

6. Coordinator of the Metropolitan Chamber, a body which brings together 21 municipalities that comprise the metropolitan city of Rio de Janeiro.

4.2 *Housing: A specific way of living. The life in conjunction with the city. Variety of uses and income*

Build, Dwell, Think like drew attention Heidegger (2015) are human operations, which can be confused with being, and with the Ente. Housing in an existence beyond the impersonality and the materialism reduction transcends the mere shelter and ends up representing what we are. The architect Rossi (1995) in his brilliant work the Architecture of the City, we also drew attention to the fact that the cities eventually generate a particular form of dwell. For Rossi (1995) each city determines a way to live, because the districts, the neighborhood, the continuous built that supports urban existence, constitute a live rail indelible form for each location. This personality of the place is afforded by the history of the development of the cluster, which is always special, and crosses moments where a diversity of sensitivities are manifest in a unique way. Thus the bélle epoche carioca, with its desire to play Paris in the tropics is a very particular manifestation of the then capital of Brazil, Rio de Janeiro. This sensitivity joins the continuous colonial, neo-classic that precedes and the continuous proto-modern, modernist, and contemporary that succeeds, generating a typology of dwell only and particular of this city.

The city of portuguese matrix, has its particular characteristics, it has always been the interdependence between the dwelling and the city, i.e., the set of facilities that support the live, such as bakeries, restaurants, sales, cafes, banks, currency, etc… In your array mediterranean, more broadly, we invariably a strong interconnection between dwell and differentiated services and uses that give support. Housing corresponds to eighty per cent of the use present in our cities, due to its characteristics of denial of a certain seasonality, the continuous housing end up determining uses and differentiated appropriations. Therefore, with a housing policy is structured can make city, or reform the pre-existing city, giving existential density to the neighborhoods, from the recognitions of its parts and different demands. In Rio de Janeiro, as in other big cities there is a diversity of parties, with diverse personalities, who end up representing particularities of the dwell. The strengthening of this particular identity should be the fundamental premise of imagination, the design and the project of the intensification of housing density in these localities. The new ventures must approaching pre stocks increasing housing densities, but respecting characteristics such as scale, texture, size in order to achieve a neighborhood with diversity.

For this, it is crucial to articulate programs for urban operations; urbanization of favelas, construction of new housing units, improvement of public space (sidewalks and roads), insertion of disabled, approximation with natural sets, etc... All these operations are retroalimentam and complement generating both the density of the dwell suggested by Heidegger (2015), as the differentiation of the various parts of the city. Here it is crucial the monitoring of public power, mainly in what refers to the value of the urban land parcels, their ideals and their possible overvaluation by its own quality. Mechanisms, already enshrined in the world, present in the Estatuto das Cidades[7], but little used by our public administrations, on account of the Iberian culture patrimonial need be used, such as; the right of preemption, the contribution of improvement, the therritory tax (IPTU) progressive, and others.

4.3 *Mobility: Hierarchy of modes. Health and mobility. Road celebration X friendly sidewalks*

The mobility in the city needs to be seen as a complex system and articulated, which aims to ensure that its entire population the broad access to all parts of its territory more quickly, and with an affordable pricing. Here it is crucial the legibility shared by all of the integrity of the system, that part of the important more high capacity until the offsets more commonplace as the sidewalks. The system should be unified allowing your access is done by a single card and universalized, ensuring that the passenger his broad access to all modes, whether they are moving objective or contemplative.

7. Law that regulated the Brazilians Cities.

The offer of systems interact strongly with variables such as value of urban land, offer opportunities for employment, leisure, culture, education, and privileging the extracts more vulnerable economically, so as to ensure that the possibility of overcoming their condition. Mobility is therefore an essential factor for promoting a better distribution of income, and therefore should be managed by the public power to reach this goal.

The disabled should include in its logic of health promotion, encouraging moves to encourage offsets that allow physical exercises, walking and pedaling on bikes shared or should not be encouraged. The adoption of loading racks of bicycles in all modes should be encouraged, without any charge, encouraging a dynamic move and promoter of health.

The use of individual car—the road celebration—should be discouraged on several fronts, such as reduction of bearing tracks on streets, widening of sidewalks and ironing crossing, revocation of the offer of car dealers in real estate ventures coming of modal high capacity, etc... The Brazilian city of Rio de Janeiro and the need to reverse its trend of massive investments in bearing races, which invariably are saturated in a short space of time. The high capacity public transport must be glamourized intending to build an awareness of civilization in the intensification of its use, gangs of movement of cars should be deleted to give rise to collective transport.

4.4 *Nature and city*

Finally, the question of the presence of continuous disasters within the great city must be stimulated so that the population is aware and approaching the dynamics of its evolution and variation throughout their cycles. In this field, the city of Rio de Janeiro has an immense natural heritage, and presents a unique situation and unusual compared to other towns, and should extend this condition in order to produce natural corridors. The presence of the Tijuca Forest, the Pedra Branca Forest, the Forest of Tinguí, of the State, the lagoons of Jacarepaguá, Rodrigo de Freitas, Piratininga and Camboriú, the bays of Guanabara and Sepetiba make this human agglomeration a single case of wild and natural metropolis. Where a series of natural continuous coexist with urban life, often relieving the tensions of everyday life, and offering leisure possibilities notables to its population.

Figure 1.

The recovery of old fountains and rivers that were channelled and hidden by the development of the big city must be exposed again and restated their neighborhoods, even involving the abolition of road constructions. The rivers are dynamic elements in our landscape, and its exposure in a humid tropical city like Rio de Janeiro are important for the population to understand the dynamics of its operation throughout the year. The rainy and dry seasons, the importance of vegetation on its banks, the fulfilment of the Marginal Protection (FMP) are factors of creation of amenities for its population, and will have a didactic value for understanding the natural cycles, uncovering much of precariousness existing today in both sanitation and solid waste collection.

The depollution of Guanabara Bay must assume a central character at this point the involvement and interaction between city and nature. The obtaining of depollution in this immense natural occurrence, of great historical value and symbolic to the city can represent for diverse neighborhoods and municipalities to obtain recovery and continuous qualified leisure. This decontamination must also deal with the unsanitary conditions of a series of bodies of water, which constitute its basin, extending the benefits of decontamination a wide territory unassisted the current metropolis.

REFERENCES

Brandão, C. 1991. A formação do homem moderno vista através da arquitetura—editora Ap Cultural 1991 Belo Horizonte. Pg 22.

Coutinho, C.N. 1970. A democracia como valor universal—acessível no site https://www.marxists.org/portugues/coutinho/1979/mes/democracia.htm FREIRE, Paulo—A pedagogia do oprimido—Editora Paz e Terra Rio de Janeiro.

Gramsci, A. 2015. Concepção dialética da história—editora Civilização Brasileira Rio de Janeiro 1966.

Heidegger, M.- Construir, Habitar, Pensar—Oficina de Artes y Ediciones Madrid.

Rolnik, R. 2015. Guerra dos Lugares: a colonização da terra e da moradia na era das finanças—Boitempo São Paulo.

Rossi, A. 1995. A arquitetura da cidade—Editora Martins Fontes São Paulo.

Rossi, A. 1977. Para uma arquitetctura de tendência, escritos 1956–1972 – Editorial Gustavo Gilli Barcelona.

Safatle, V. 2009. Só mais um esforço—Três Estrelas São Paulo 2017 SINGER, André—Os sentidos do lulismo, reforma gradual e pacto conservador—Editora Companhia das Letras São Paulo.

Tafuri, M. 1981. Teorias e História da Arquitetura—Editora Presencial Lisboa.

Tafuri, M. 1987. The Sphere and the Labyrinth, avant-gardes and architecture from Piranesi to the 1970,s—MIT Press Bosto.

Tota, A.P. 2009. Os Americanos—Editora Contexto São Paulo.

Drawing (...) City (...) Body, Dwelling on Earth – Janeiro (Ed.)
© 2019 Taylor & Francis Group, London, ISBN 978-0-367-02602-8

Possibilities of the flesh of the body and the flesh of the things: The drawing as a conversion of the *visible into an other-visible*

P. Janeiro
CIAUD, Lisbon School of Architecture, Universidade Lisboa, Lisbon, Portugal

ABSTRACT: I live in a body and want (or, desire?) to draw that thing. It seems as if the thing is external to me… Yet still I say that I see it…

That thing, that or it which is not me or I am not; that thing which I know is not my body, while flesh of my body, is there; therefore, if it is there it is because, in flesh, it doesn't belong to me while consubstantiated with me, in-me, or, belonging to the flesh of my body.

Keywords: Drawing, body, architecture

1 INTRODUCTION

As a principle, this isn't totally true, because: I'm also capable of seeing, for example, my hands, my own body almost to its entirety, my whole body as a reflection in a mirror…

But there we are…

However, instead of talking about the drawing which is summoned by things (apparently external to me, and to my body's flesh), should I rather start by reflecting on the drawing which depicts what I see as being my own flesh, the drawing of my own body?

Could it be, that, instead of talking about things, or the drawing of those things, which don't incarnate me in substance, I should instead be talking about the drawing of the movement of my fingers while I type this text on the keyboard of a computer, or better still, be talking about the drawing of that which is visible to me of me in the present scenery where I find myself in, or, finally, draw myself whilst submitting to my own reflection on a mirror…

I choose not to talk about it—the events which I may observe regarding the flesh which is my own body.

I decide to talk about the Drawing of things which are, apparently, external to me—the things—the things seen from this side of the mirrors; the things outside of myself(?).

I choose to talk solely about that which is observable by me, amputating myself in body, as if the flesh of my body existed only from my eyes onward: I will only speak about the observable event, the phainomenon, of that which can be seen, of the phenomenon.

I do not wish to talk about my body, even though he is implicit to me—being that it is he who sees; in all instances it is he who draws me or draws himself.

One cannot escape the body.

I speak of observable events—I speak of visible things in front of my body; and, unwillingly, I speak of my body which is, after all, who observes and draws them.

(I could also speak of invisible things. Invisible worlds, invisible architectures, invisible cities which might or could have happened: Filarete's Sforzinda; Francesco Colonna's Hypnerotomachia Poliphili, or other dreamt cities; the 54 cities in Thomas Morus' Utopia; Ebenezer Howard's Garden Cities of To-morrow; Lloyd Wright's Broadacre City; Buckminster Fuller's Cloud Nine; all invisible in flesh; or other cities yet to happen…

All of them, once mentionable by us, are therefore existent in drawing; but, I shall not speak of that. Neither will I speak of the invisible.)

Rather, I will speak of the sensitive event, the phainomenon, the phenomenon; however, I will, within the typologies of phenomena, be speaking of the observable phenomenon, the thing which can be seen.

Phenomenon appears, for example, in the work of Kant (1997), as all that which is object of possible experience, characterised by form and matter, understanding representation as an act of subjective freedom—the problem poses itself unto a subject via a moral point of view (Kant, 1997).

On the other hand, from a phenomenological point of view, Husserl (1913, 1962, 1999, 1991, 1989) and Merleau-Ponty (1999), both clarify that the phenomenon is object of intuition or of immediate awareness, being at the same time a manifestation of essence.

Contrarily to Kant, who considered the phenomenon to be a sensitive manifestation, "the undetermined object of empirical intuition" (Kant, 1997, p. 61) – "subdued to the formal conditions of space and time" (Kant, 1997, p. 61, 62) –, of a definitively unaccessible thing in itself, Phenomenology states that things themselves are revealing, and that the subject, "source of sense and intentionality" (Lyotard, 1999, p. 36), is understood as one to whom (or, in whom) those things reveal themselves.

The phenomenological project "renders possible the reconciliation of objectivism and subjectivism, of abstract knowledge and concrete life" (Lyotard, 1999, p. 39), thus letting the world—on this side of the scientific border—unveil itself by means of immediate intuition and concrete experience, or, in other words, a world which is situated the closest to us, rather than the objectivism which threw the world into a "logical-mathematical formalisation" (Lyotard, 1999, p. 38) and the "mathematization of knowledge" (Lyotard, 1999, p. 39), sustaining that, in the phenomenon, the vision of essentiality in things is made possible due to a return to the originality of those same things. Thus—as we arrive to the question of originality—the reference to Phenomenology is of particular pertinence to the topic we are discussing.

What originality? To which origin are referring to? Let us suspend these questions for now.

Only upon the phenomenon may we build knowledge, never upon the vague impression that things are more saturated than what we are capable of reading from them. They are, what they are to me, in-relation to myself, as entities that manifest in me at a certain moment and not, "against the transcendence of kantian in-itself as a product of a philosophy of understanding [Cf. the notion of Kant's "understanding" in Critique of Pure Reason, when he says "Generally speaking, understanding is the faculty of knowledge. This consists of a determined relation of representations given to an object." (Kant, 1997, p. 136)], to which the presence of the object is no more than a simple appearance of a hidden reality" (Lyotard, 1999, p. 45) – furthermore, that is what we learn from Nietzsche when he says: "Underneath this reality in which we exist and live [...], a very different one is hidden [...]" (Nietzsche, 2002, p. 40, 41).

From this point of view, things are only that—they exist to the subject as evidence: "We truly are, and the evidence is 'the experience of truthfulness'." (Merleau-Ponty, 1999, p. 14). Furthermore, it is as evidence that things may be experienced. Therefore, no other reality of things exists. Subsequently, to consider a transcendence of things implies the consideration that those same things are somewhat superior in relation to the subject, that they somehow surpass the subject. That would admit an own nature of things, an immanence of the object to consciousness. In fact, this transcendence, which we may call vague impression, or "vague solicitation" (Merleau-Ponty, 1999, p. 288), when referring ourselves to the need of synchronicity between subject and constituted reality (which is in him, and in himself, within reality), before characterising an outer to the world, the own intentionality of the subject has inscribed the heart of that world, thus, this capacity that the conscience has to relate to that which is not, to tend to a somewhere, an outside of itself, a sort of non-place.

On the other hand, we have already intuited that the subject constitutes a representational reality. He represents a reality in which he inscribes himself, better still, the subject inscribes that reality in himself (in the subject), by means of representations, and in doing so he constitutes it, eventually, as transcendent. Well, he is the drawer.

Accordingly, the question on the transcendence of things is no longer posed on wether they, by themselves, had a will, a desire or an intention of existing, after all, "the notion of phenomenology itself deals with that question: conscience is always a conscience of, and there is no object which isn't an object to" (Lyotard, 1999, p. 43); in other words, and reaffirming our conviction, the object exists only in relation, and the conscience is that to whom those relations exist, and which, consequently, without that possibility of relation is-not.

Intentionality, wherein a phenomenological approach might lie (more: which "finds itself in the center of phenomenological thought" (Lyotard, 1999, p. 55)) –, is the subject's own thing—so defends Husserl when he says: "Intentionality, as a fundamental property of my own psychic life, designates a peculiarity that belongs to me whilst man [...]" (Husserl, 1992, p. 42), it is the activity by means of which the conscience relates with what it aims for. The conscience aims for what it feels, and if it feels itself feeling, it aims at itself, therefore intentionality is an objective, a donation of sense (in the "interlacing with the world" (Husserl, 1992, p. 37) through perception), and not a thinking thing as Descartes (1997) defined it in his Discourse of Method the following manner, "[...] this I, that is to say, the soul, by which I am what I am, is entirely distinct from the body that, even if it were to be there no longer, the soul could not help being everything it is." (Descartes, 1997, p. 36), but which Merleau-Ponty, as if unveiling a secret, says:

"The world which I distinguished from my own self as being the sum of things or processes linked by causality, is rediscovered by myself 'in me' as being a permanent horizon of all my cogitations and as a dimension in relation to which I unceasingly situate myself. The true Cogito doesn't define the existence of the subject by his thought of existing, it doesn't convert the certainty of the world into the certainty of the thought of the world, neither does it substitute the world itself by the signification world. On the contrary, it recognises, my own thought as as inalienable fact and eliminates any sort of idealism revealing myself as a 'being in the world'." (Merleau-Ponty, 1999, p. 9)

Now, this is exactly what the Drawing is: a visible manifestation of that intentionality which mobilizes the body; not only does it mobilize, it impels the body to register the instant in which that mobilisation occurs. It's in this sense that Almada Negreiros says "the drawing is our understanding establishing that instant" (Almada Negreiros, 2006, p. 4).

The body transcending itself into Drawing; into pure transcendence: in that mobilisation, in that impetus and in that action whilst drawing.

Being pure transcendence, the conscience, never coincides with itself and is fundamentally temporality, openness to both past and future. Intentionality implies at the same time the signification—by which it is possible to represent –, that is the overcoming of a simple fact, that of overcoming the ascertainment of the occurrence of a form.

On one hand, due to the fact that things do not feel, and, because they are there in front of the subject as evidences, they aren't, therefore, for themselves, and by not being, for themselves, they don't exist for them, they exist solely for the subject—being whom can think them—and thus, consequently, they are no more than provisional representations, of built images, of entities that possess signification, that is to say, they only exist because the subject has summoned them to existence while determining them as a formal occurrence whose model is an interpretation of a fact from a certain point of view. On the other hand, it might have been a conscience that things are there, a certain conscience of the distance (purely extension) which separates the subject from the things, which could, apparently, argument a certain objectivist standpoint which advocates the existence of a interiority (I) and an exteriority (the world). This standpoint should be eliminated (at least from our point of view) because the intentionality (as the manner by which the conscience relates to that which is seen) is an objective, that being a donation of sense, relativises (by interlacing with the world through perception) that distance, let us say, physical, which may be translated by the measure; and in doing so, brings the I and the World, combining them.

And because this combining is all the more evident between the subject/inhabitant and the architectural-object/habitat. It shall, hypothetically, be the most evident of subject/object relations, further along we shall see how and why.

The things manifest themselves to the drawing subject as evidences; it is the drawer who constitutes them as things which, eventually, transcend him.

However, the evidence is, from the start and at least, a certainty—that the object is there. In a certain way, it manifests myself as a presence in front of me. And if it is in front of me it is because I, from my point of view, detect it.

There is, in fact, a a physical distance between me and it: a distance, shall we say mathematical and measurable, between the subject and the object; however, another distance exists, one which would be difficult to quantify (perhaps the metaphor of the "precipice" (Heidegger, 1988, p. 35), might be adequate here)—because the object as a physical-chemical substance doesn't belong to me, it or I do not share the same flesh, the same-body. However, something is shared in this relation: that something is our mutual presence—in the now (that is where time appears to me as a consequence of that sharing; where time is born, because the conscience of time which appears in the constitutive relation of the subject upon the object—through which the subject represents it and by which the subject may have conscience of himself—is a synthesis of identity; and, where the object is a connecting thread). In a certain way, we are contemporary, at least, to that moment of time when the detection takes place. That moment, that conscience of our coinciding, I-here and it-there, better still, it-there-in-I-here, the concrete certainty of it, in that movement (whereby perception brings the object to me and suppresses the precipice) which is experienced in this actual experience (and that "may be defined as a lived experience of truth" (Lyotard, 1999, p. 39).

The ascertainment that the object finds itself at a certain distance implies that the subject of that object knows that he doesn't possess it; that, in a certain way, the object is beyond him. This is a state of things we can not deny. After all, they, both subject and object, do not partake of the same flesh. Therefore, we must observer this relation from another point of view.

It's evident that the relation established between the drawing subject and the drawn object mustn't be placed by the point of view of the substance which is their support: on one hand, the support of the subjects' existence as I (within a body), and the other hand, the support of the objects' presence as thing (within a form). However, if we all agree, that in spite of them not being supported by the same substance, a relation exists between the subjects and the objects that surround him, that is because those things arrive to him by another vehicle which is not that of biological substance.

To continue with our reasoning: if the subject perceives the object it is because the latter manifests itself unto him. However, an analysis to that manifestation, which we previously named as evidence (that certainty that the object is present, thereby, in front of) may only happen if we proceed to it from (and, to) the subjects' point of view, because the object, by itself, is not distinguishable from the subject and is ignorant to the manner by which this appearance happens to unveil its' properties.

Let us go a few steps back: the object is in front of the subject and there is a physical and/or biological distance that separates them. This distance, however, is irrelevant in the exploration of how it happens, related to the manifestation or evidence. In this analysis, distance is merely instrumental. Nonetheless, let us continue: the notion that we are not supported by the same substance, the notion that I "am not" the object, nor the object is I, means, in short, the notion that we are separated in space, implies the recognition that, in fact and above all, there is a distance which can't be ignored. It is by being conscious of this distance, determined by our positions in space, that the subject may judge that-which-manifests-itself. Or, better still, it is simultaneously, through recognising that distance that he may distinguish the places, the centres within that relation: the place from where he sees and the place of the seen thing. Therefore, the subject recognises: the I who is always here and, from here, the object which may always be there (the place where the thing may manifest itself). The I who is always here is the body, he is the field where my perceptive powers are located, he is where (or, from who) things manifest themselves, where things may have. At heart, the body is the absolute place where "appearance is reality, [where] the consciousness' being is to manifest itself" (Merleau-Ponty, 1999, p. 504). In short, the body "is the vinculum between the I and the things" (Merleau-Ponty, 1991, p. 183).

But, if that is true, if the body really is that vinculum, will it make sense to even speak about the distance which separates the subject from his object? If, in fact, as Merleau-Ponty argues, perception is the "background from which all acts stand out and is presupposed by them" (Merleau-Ponty, 1999, p. 141), and without it, therefore, we know not of the world, thus, does it make any sense to even talk about the distance which separates the subject from his object?

If we all concur, the drawing subject and the drawn object do not share the same flesh, for there is another kind of sharing between them, there exists another relation, one which cannot be explained by Biology or Physics.

Indeed, we must all agree, that it is the body that brings himself the world. The subject resides in the body and it is through him that the world arrives to the subject.

The distance between the subject and the object reveals itself as only one more variable in their relation, because, "the relation between it [the body] and them [the things, the object (generically)] is that of an absolute here with the there, of the origin of distances to distance" (Merleau-Ponty, 1991, p. 183)

Therefore, not only does distance appear with the body and within it, but also the object appears with the body and within it. And if we owe this appearance to the body, then, at least, we must admit the hypothesis that in the subject/object relation, what exists is an assemblage of the body to his object, or in other words, a combining of both thing that feels to felt thing.

After all, one must draw if the body asks to. This is why there are no right nor wrong drawings. The theme of Figurativity in Drawing is another theme.

REFERENCES

Heidegger, M. 1988. A Essência do Fundamento. Lisboa: Edições 70.

Husserl, E. 1913. A Filosofia como Ciência de Rigor, Primeiro Tomo (Ideias Directoras para uma Fenomenologia Pura e Uma Filosofia Fenomenológica). Lisboa: Ed 70.

Husserl, E. 1962. Expérience et Jugement. Paris: Presses Universitaires de France.

Husserl, E. 1989. La Terre ne se Meut Pas. Paris: Ed. Minuit.

Husserl, E. 1992. Conferências de Paris. Lisboa: Edições 70.

Husserl, E. 1992. Méditations Cartésiennes, Introduction à la Phénoménologie. Paris: Libraire Philosophique J. Vrin.

Husserl, E. 1999. L'Origine de la Géométrie. Paris: Presses Universitaires de France.

Kant, I. 1997. Crítica da Razão Pura, 4ª ed. Lisboa: Ed. Fundação Calouste Gulbenkian.

Lyotard, J-F. 1999. A Fenomenologia. Lisboa, Edições 70.

Merleau-Ponty, M. 1991. Signos, 1ª ed. São Paulo: Martins Fontes.

Merleau-Ponty, M. 1999. Fenomenologia da Percepção, 2ª ed. São Paulo: Martins Fontes.

Nietzsche, F. 2002. A Origem da Tragédia, 11ª ed. Lisboa: Guimarães Editores, 2002.

Drawing (…) City (…) Body, Dwelling on Earth – Janeiro (Ed.)
© 2019 Taylor & Francis Group, London, ISBN 978-0-367-02602-8

Possibilities of overcoming aesthetic preconceptions in drawing learning: Exercise notes about sensitive impressions

R. Ferreira Lopes
Department of Design, Representation and Technology, Faculty of Architecture and Urbanism, Federal University of Juiz de Fora, Juiz de Fora, Brazil

L. Mendes de Vasconcellos
Postgraduate Program in Architecture and Urbanism, School of Architecture and Urbanism, Fluminense Federal University, Niterói, Brazil

ABSTRACT: This article aims to present some considerations about the use of other perceptive modalities in the teaching of observational drawing. In view of the resistance of a good part of the students enrolled in the courses of Architecture and Urbanism to the drawing learning, it was assumed that, in the teaching of this practice, it is necessary to overcome preconceptions that lead students to a risky belief that the ideal is to produce a graphical representation only committed with the verisimilitude of visual images. In order to investigate this assumption, some didactic experiments were accomplished in the elective course Urban Sketches of the Design, Representation and Technology Department of the Faculty of Architecture and Urbanism of the Federal University of Juiz de Fora (DPRT-FAU-UFJF), held in the year 2016. Under a Phenomenological approach, the discipline was an alternative to traditional drawing teaching. Thus, the methodology used includes a theoretical revision guided by the Phenomenological current of thought, based on the concept of embodied image; as well as an empirical description of the performed activities in the second class of the discipline, entitled "Overcoming traditional preconceptions". Based on the statements made by students in class, as well as some images extracted from these exercises, it was confirmed that the results expected for this research respond adequately to the raised hypothesis.

Keywords: Teaching, Observational drawing, Embodied image

1 INTRODUCTION

The purpose of this paper is to present some considerations about the use of other perceptive modalities in the teaching of observational drawing, in view of the resistance of a good part of the students enrolled in the Architecture and Urbanism courses to the learning of this competency.

Based on the doctoral thesis of Lopes (2017), it is understood here that the thought and behavior of the new generations of university students has undergone a significant change in recent years, since these are influenced by an increasingly technological and imagistic culture. The concern expressed by Lopes (Ibid.) corroborates the problematic of Pallasmaa (2011), when the latter associates the predominance of the vision in relation to the other human senses, including the mode how architecture is conceived, taught and criticized. For the theorist, contemporaneity brings with it an enormous load of technological excesses and sensorial stimuli condensed into a symbiotic construction with visuality. In this perspective, the only sense that is fast enough to keep up with the astonishing rise of the speed of the technological world is vision.

Arnheim (2012) states that we have neglected our "gift" of understanding things through our senses. Our eyes, therefore, have been reduced to instruments for identifying and measuring, and indeed we have suffered from a certain lack of expressible ideas in images and a "capacity to discover meanings in what we see" (Ibid., p. XIII). A paradox emerges: if, on the one hand, our contemporary culture is predominantly visual, on the other hand, the teaching and practice of drawing has lost its value, especially when it comes to the development of visuographic skills. In more than a decade of teaching in higher education, the present author has ascertained in many students enrolled in artistic disciplines, profound psychological and behavioral effects resulting from this loss of references.

In the aesthetic scope, it is also noted the establishment of certain preconceptions, manifested in the search for an "ideal of perfection" of what "there it is" being seen. The predominant desire to represent a "beautiful" image—perhaps a remnant of the academic traditions of Beaux-Arts still present in our culture—carries with it the stereotyped commitment to keep the observational drawing accurate and true to the visual reality: a way of imitating nature. In this sense, the image is conceptualized as a substitute of the real object, associating the myth that to draw in a satisfactory way it is necessary to have "special innate gift". It becomes imperative, in students discourses, the need to use digital technologies to ensure such results.

With the speed of technology and the predominance of visual sensation in our culture and society, the acquisition of knowledge is being replaced by the superficiality of immediate information, and this also extends to the teaching and learning of drawing. The change in the representational paradigm implies in new attitudes in the educational scope. As an alternative to the teaching of traditional observational drawing, the didactic strategy in question was based on the hypothesis that learning from a perspective of the sensitive body is precisely the cognitive gain obtained through the experience of the concrete world. It is understood that this gain does not occur in the sense of passively representing reality in a mimetic way to the intrinsic qualities to the spatial form, but as a way of letting the consciousness-body act and produce knowledge from the contemplation of things as they are taken by senses.

Therefore, in order for the teaching of observational drawing make sense to students, it is necessary to overcome the preconceptions that lead students to a risky belief that the ideal is to produce a graphical representation uniquely compromised with verisimilitude, i.e. with the representative trustworthiness that accompanies the referring image. Based on this assumption, there were some didactic experiences in the elective course Urban Sketches (DPRT-FAU-UFJF). Offered in two academic semesters of the year 2016, the discipline, under a Phenomenological approach, was an alternative to the traditional, purely visual, Representationalist approach, commonly used in observational drawing disciplines.

Thus, the methodology used in this paper includes the following steps: (i) theoretical review based on the Phenomenological concept of "embodied image" (Merleau-Ponty, 1971; Pallasmaa, 2013), in opposition to the Representationalist conception of the world; (ii) empirical description of the activities accomplished in the second class of the discipline Urban Sketches, entitled: "Overcoming traditional preconceptions". In the description of the experiences are some observations that support the didactic strategies proposed in the discipline, as well as the theoretical and conceptual elaborations. Lastly, a synthesis of the significant points seized with the students in the mentioned meeting is established in the discussions. Based on the statements made by students in class, as well as some images extracted from their didactic exercises, it was confirmed that the results expected for this research adequately respond to the hypothesis initially proposed.

2 ANOTHER LOOK: THE OBSERVATIONAL DRAWING UNDER THE BODY'S PERSPECTIVE

Visibility is the essence, by nature, of the observational drawing. The teaching of drawing, based on the graphic representation of the visual space of immediate perception, instrumentalizes the student to describe the perceived reality. The image resulting from an effective visual impression on the individual about the apprehended psychophysiological space, becomes the final object of artistic expression. Showing things as we see them in perception defines the character of "representation" of the drawings, because the monocular traces, developed since the theories of perspective from the Renaissance, have a close relationship with the artistic techniques of the observational drawing and painting, to reproduce satisfactorily the depth effect in a two-dimensional picture (Mueller, 1977).

This thought reflects the Representationalist conception, which understands that we are separated from the world and that it exists apart of our experience, thus causing the segregation between the subject and the object. Indeed, in art, perspective presents as a mental model the fragmentation between the observer subject and the observed object by means of an imaginary plane, i.e. the perspective picture, whereby it is projected the representation of the world in an attempt of dimensional construction of the space. The Representationalist theory, epistemological mark still prevailing in our culture, understands that the brain passively receives external information already ready (Mariotti, citated in Maturana and Varela, 2001). Representationism suggests that knowledge is a phenomenon based on the mental representations we make of the world. The mind acts as a "mirror of nature", since the world contains information and man's task would be "to extract it through cognition" (Ibid., p. 8).

In our daily lives, by imposing cultural conditions, we tend to live in a world of certainties, in which things "are what they are" because that is how we "see them". This culture is reflected in the teaching of drawing, because, on the one hand, teachers demand from the students the correct graphic description of reality, where the representations must present satisfactory figurative qualities, stopping to the full correspondence of the being with the referent in the representation of dimensional space. On the other hand, the students establish certain preconceptions that lead to the preoccupation with the elaboration of a mimetic observational drawing to the real, reverberating in the commitment to "live in certainty".

In another sense, different from what is assumed in Representationalist thought, the notion emerges that to perceive is not a passive relation to the environment. For Ostrower (2008), perceiving is an activity, it is intentional. In this understanding, drawing is not a passive activity, because who draws start inhabiting it in a symbiotic connection; a means of fertilization that crosses the artistic competences with the intellect. With an integrally engaged conscience, those who draws can merge with the image, because together they create and share the "conscious being of drawing", incorporating in this relation their body, perception, imagination and memory. The experience of drawing integrates us phenomenologically by the act of meaning itself, molding us and allowing us to be molded by drawing; to inhabit it when we leave ourselves for a world that has meaning for us. It is understood that, in the effective action of the students, the subjective consciousness and the world are not separated by a picture, but closely interlinked, because the drawing graphically represents the world perceived from a living-observer in its interactions with the environment. For this, it is understood that the student can be involved in the experience of the drawing in an active way, not being therefore unrelated and passive to the world in which we exist.

This thought converges with the conception of Merleau-Ponty (1971) in his work, Phenomenology of Perception, 1945. For the philosopher, perception unites the activity of thinking to the activity of feeling, making the human body the center of the world of experiences. In this sense, Merleau-Ponty's studies focus on perception in general and on vision in particular, however, instead of the objective and external eye, the "Merleau-Ponty's sense of vision is a embodied view" (Pallasmaa, 2011, p. 20). Merleau-Ponty understands the experience of the body as the creative field of the senses, through different views of the world, because perception can not be understood as a mental representation, but an event of corporeity and thus of existence, i.e. of the "lived-body".

The phenomenal body is spatial, not in the sense that the objective body is "in" space, but in the sense that "our primordial encounter with being and that being is synonymous with being situated" (Merleau-Ponty, 1971, p. 258) and oriented in the experiential space, thus relativizing the Cartesian meaning of space. So, existence itself is spatial, since spatiality is "the primordial condition of living perception" (Merleau-Ponty, 1994, citated in Santaella, 2012, p. 23). For Merleau-Ponty, the geometrically idealized objective space is distinct from the existential, anthropological or lived space, since what is depth for an observer, may be width for another whose look is oriented perpendicular to the first. When the individual is located and naturally oriented it presupposes a bodily abode in the space experienced by him, for true depth is given to a living perception that inhabits space. Not only are sensory modalities such as seeing, hearing, touching, smelling, and proving mutually influencing, but they are influenced by affective and emotional responses to things, since the perceptual experience is experienced by a perceiver.

The concept of "embodied image", which Pallasmaa addresses in this same theoretical matrix, emerges as "a specialized, materialized and multisensorial lived experience" (2013, p. 11), where, simultaneously with poetic images, "evokes an imaginative reality and become part of our existential experience" (Ibid.). The embodied images and poetic images therefore play a decisive role in the mental world of the individual. The perception of an image for Merleau-Ponty is an integrated experience, in which one perceives with the whole being, a unique structure that communicates in a single stroke, all the senses, evoking in an absolute way the lived reality. The concept of "multisensory image" presented by Pallasmaa (2013) emerges from the Merleau-Ponty's conceptions that the visual image mediates with other sensory experiences that "could even dominate the nature of the image" (Ibid., p. 53). Thus, Pallasma emphasizes the sensorial interchange that occurs in a profound way in human perceptions, highlighting the innate tendency of these connections to merge in their totality in the sensorial experience.

From the Phenomenological point of view, the proposition of a teaching of drawing that encompasses the multiple senses of the individual means to construct his vision based on the consciousness and the sensitivity of the being itself. In this sense, the visual image seized and described in the exercise of the drawing can establish a mediation with other sensorial experiences. The production of multisensory images has the purpose of allowing the students to identify with the drawing through graphical and textual annotations taken by the senses, enabling the recording of the description of sensory data of the perceptive phenomenon during the drawing exercise. Drawing does not mean to freeze an image in a photographic instant, but to spend time observing, letting it affect or impregnate itself in the experiential space-time; mixture between cognitive and affective plans related to the body in the sensitive experience of the world. Relationships in the lived world and in information processing occur as embodied and sensory activities, not purely visual concepts or awareness data. The production of multisensory images was the main thread for the teaching strategy from a body perspective, as explained below.

3 DESCRIPTION OF THE EXPERIENCE: "OVERCOMING TRADITIONAL PRECONCEPTIONS"

The Urban Sketches discipline (DPRT-FAU-UFJF) was composed of a total of fifteen meetings. Practical activities were proposed as a way to bring students closer to the architectural and urban experience, initially taking as main cut-off the city of Juiz de Fora-MG. The purpose of the second class of the discipline, named "Overcoming traditional preconceptions" was to put in "parentheses" many preconceptions arising from Representationalism, demanding for this the suspension of conventions of the traditional perspective method.

To draw the objects of perception some laws of the traditional perspective have been disregarded as vanishing point, since it does not manifest in our visual field. There is only a sense of direction that signals the parallelism of the contour lines, which in its turn represents the threshold between the figure and the background, or two planes competing with each other. The students were instructed to delineate the contours of the depth, with greater artistic and gestural freedom, by the synergistic and intentional relation of the visual sweep with the synchronous movement of the hand.

Likewise, the imaginary horizon line levels by definition at the level of the observer's eyes. This representation falls to the ground when it is understood that the threshold between the earth's surface and the celestial vault is curved and does not always appear clear before it.

However, in understanding that other individuals, when they position themselves on the same level as the observer, reveal their own horizon; then the other can be recognized by seeing him as equal, taking his eyes as if they were those of the proper observer, dialoguing with them and sharing his experiences of the world. It means that observing the other requires observing oneself while observing the world, because between the human being and the world there is no hierarchy, no segregating plan, but a constant cooperation in movement involving embodied intentions ("I can") and the sensory-motor skills that occur in this dynamic. This active conception breaks with the Representationalist notion that we passively receive outer information and commands and the disembodied mentalization of represented things ("I think"). The active relationship of the body to the environment triggers the action that makes the being and the world inseparable. In this sense, to overcome the preconception of the predominance of visual images over the senses, it is reflected on the possibility of "turning off" the visual sense in the perception, attending to the tactile and gustatory senses.

Before the main activity planned for the class, the students were asked in the first exercise brushstrokes that guide the fluidity of the paint and the need to master it with skill and speed, in order to explore the possibilities of stains and the unpredictability of the results, in substitution to the contour lines in the perception of form. The exercise consisted in "feeling" the brush slide through the paper, in successive overlapping layers and in scales of tonal gradients (Fig. 1).

For the second exercise, numbered opaque paper bags were distributed, containing a chewable candy. The guests were asked to list their drawings according to the given reference and with their eyes closed they would throw the candy in the mouth so that, without biting or swallowing it, they could slide their tongues over it, collide it in the palatine veil, savour their tastes, and describe all of these experiences on paper (Fig. 2). Several forms appeared in the description, without the concern of comparing the graphic referent with what was truly identical to the perceived entity, since the visual forms of these candies were not revealed. This experience has shown that the movement of the candy in the mouth and the perception of the palate can give us the notion of space. The candy exercise revealed to them in principle that the palate involves touch and smell as the tongue moves around the candy, exploring tactitely its shape and its peculiarities (texture, elasticity, hardness, porosity, smoothness, granularity, etc.). In addition, the smell registers the aroma, sweet, bitter or sour. Many students reported having the impression of "feeling the color" in the candy's exploration, associating color with the aromatic and gustatory qualities of all sorts of fruity flavors provided.

The third exercise, in sequence, developed the tactile experience from the blind drawing. Six different solids were distributed, each wrapped in a closed, colored paper bag that corresponded to the reference number of that type of surface. Three copies of each of the six types were produced totaling 18 (eighteen) bags, which were delivered to each student who, in turn, had fifteen minutes to grope with his hands and draw the shape on the paper, verbally describing at the same time, their sensorial qualities. At the end of the time, the student passed the bag

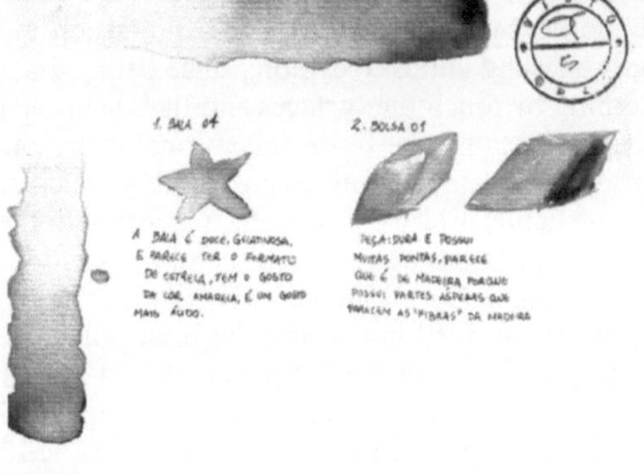

Figure 1. Fluid and watery brushstrokes on the paper in the painting of the student LSRT. Source: Author's photograph, 2 Sep. 2016.

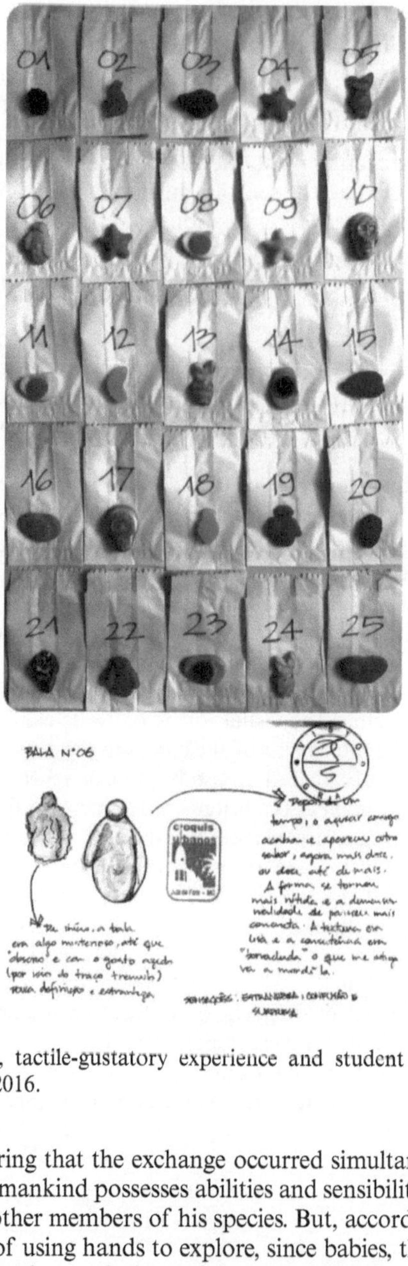

Figure 2. Assorted candies, tactile-gustatory experience and student DRC's description. Source: Author's photograph, 2 Sep. 2016.

to the next colleague, ensuring that the exchange occurred simultaneously, keeping a circular flow (Fig. 3). It is true that mankind possesses abilities and sensibilities and uses their hands to know, caress and comfort other members of his species. But, according to Piaget and Inhelder (1993), it is innate the gift of using hands to explore, since babies, the physical environment—differentiating and discerning by touch the surfaces and their shape. The rupture with the traditional conception occurred because the three-dimensional experience was not provided by the perspective of the vision, but by the sensations mediated by touch and kinesthesia when the movement of objects by the hands is also considered. According to the students' descriptions, touching and manipulating mysterious objects with their hands produced in the imaginary "a world of things" due to the complex forms meticulously modeled for the exercise (Fig. 4).

For strategic reasons, it was decided not to use the basic solids (cube, cylinder, cone and sphere), since it would facilitate the recognition of pre-conceived shapes in the imaginary. Thus, six solids with non-Euclidean topological surfaces were used, namely: (i) Regular Hexa Hyperbolic Paraboloid; (ii) Steiner's Roman Surface; (iii) Klein Bottle; (iv) Möbius Strip; (v) Plucker's Conoid; (vi) Parabolic Conoid. The proposal aimed to detach from the representation of shapes preconfigured in the imaginary, as basic solids or polyhedra (also known as Platonic solids).

Figure 3. Student describing her tactile experiences. Source: Author's photo, 2 Sep. 2016.

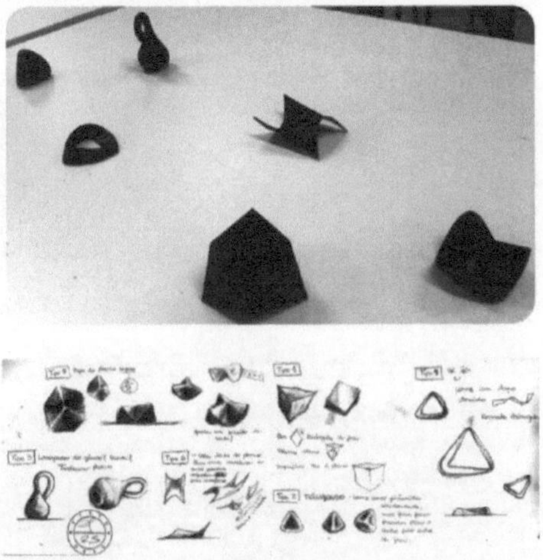

Figure 4. Topological solids made with 3D printer and the graphic description of the student PAC. Source: Author's photograph, 29 Apr. 2016.

After five rounds, the last object was drawn totally blindly, so as not to divert attention, and with one hand in the drawing and the other inside the paper bag.

4 DISCUSSION

According to the reports of the students collected by the present author in class, the above experiments confirmed the initial hypothesis that direct experience goes beyond the purely visual description and that to feel personally involved there's no need to worry about what we see, which reaffirms itself as the great myth of representing the world with absolute certainty and full fidelity.

The idea initially proposed for the exercise was that the spatial and flat geometric figures would allow the recognition, organization and mental systematization for the later graphical representation on paper. However, the topological surfaces generated some mental confusion in the students due to the way in which the things were apprehended in the tactile perception. According to their perceptions, it is concluded that this difficulty was manifested by the fact that there was no previous experience with form, nor because there was no clear distinction between figure and background, making the contours lose their meaning. The kinesthetic movement also allowed a global notion of being in various positions of space at the same time, without a fixed perspective of the object or a Cartesian orientation of space.

According to the description of one of the students during the exercise: it surfaces "something that is not understood mentally, but that is directly felt in the experience with the hands" (testimony of the student LDBSN, 2nd class, 2 Sep. 2016). Therefore, it was understood that the imaginary allowed the students to evoke associations that split with the projection of memories. The proof of this observation was given by the description of a student who compared the surface of Plucker's Conoid with Jørn Utzon's architecture for the bark of Sydney Opera (1973) (LEAS, 1st class, 29 Apr. 2016) (Fig. 5) and a student (HBS, 1st class, 29 Apr. 2016) who associated the Roca Gallery in London by architect Zaha Hadid (2011) with the Parabolic Conoid.

Without the notion of contours, reinforced by the absence of the figure-background and of some solids which by the nature of their own shape have no obvious contours, the direct appli-

Figure 5. At the top: "Plucker's Conoid", complex figure modeled in 3D printing for exercise; in the middle: the Sydney Opera House, drawing by the author; below: student LEAS drawing (29 Apr. 2016), who associated the touched object with the Sydney Opera House. Source: Author's collection.

cation of ink water on paper proved more effective. Loose brushstrokes revealed spots that pictorially represented the masses felt by touch, keeping the synchronicity between affectation and graphic narrative, hardly reached in the graphics of construction lines in the drawing.

Some students revealed that they "felt the color", sometimes associating with the color reflection of the paper bag, or imagining the materiality of the felt objects, printed in ABS plastic (described as being white—mistakenly, once the objects were black plastic. A student felt "lost in space, traveling and not arriving in the same place" (testimony of student ALO, 2nd class, 2 Sep. 2016), because many objects "had no edge," and no vertex and thus did not allow the Cartesian conception of the point of origin or any other reference than the contact of the body with the perceived thing.

Other students revealed that they felt agony for not viewing the objects. Others have already felt a strong desire to take the model to the mouth, to bite and to feel at the same time. Finally, in the experience of totally blind drawing, the students claimed that they perceived better the felt shape, synthesizing the graphic representation in general lines and a safer trait than the previous ones.

5 CONCLUSION

The didactic experiments accomplished in the Urban Sketches discipline demonstrated that the exercise of multisensory image drawing can contribute to a description of the world observed and imagined by the student who draws, thus allowing an investigation of the being, his past, his memory and his imagination. Such experiences have also allowed the understanding that the poetic and plastic dimension of the body generates other possibilities of reading the real and the sensitive language. Therefore, the activity of drawing is placed as a field of possibilities for the experience of the sensible, not only with the thought of seeing/looking, but also as body reflection.

It is further believed that the student of drawing should not be conceived as merely a "mechanical recorder" of the surrounding visual reality, just as one may regard the artistic representation of an object as a tedious transcription of minutiae of its visible appearance. Here the images of reality, full of meanings and immersed in a field of phenomena, have validity even if they are distanced from "realistic verisimilitude", agreed by academic norms and Representationalist thought.

By freely assuming the attitude of overcoming preconceptions ("I can"), or at least suspending them without detracting from their purposes, students were able to free themselves from the blockage of the prevailing culture ("I think"). So, the search for abstract geometric entities that mimetically represent the world "out there" and the belief in digital technologies as the only solution to overcome the representational problem, are also in suspense.

When verifying in the student entering the course of Architecture and Urbanism the behavior of denial of drawing, disbelief, anguish and even fear, perhaps still imprisoned by the commitment of verisimilitude, it is necessary to seek the syntony between the student and the world during the activities proposed in a discipline of drawing. If in the everyday existence the incoming students have their lives mediated by the technology experiencing the time like a fast and intense flow, naturally, when they come across the drawing, tend to be taken by some feeling of "displeasure". In this perspective the class becomes tedious, it happens slowly, expressing the impatience. Therefore, the use of the explicit multisensory experiences aims to sensitize the dimensions of the multiple perceptive modalities, thus extending the horizon of visibility and reaffirming the personal identity of the student through graphic expression.

ACKNOWLEDGEMENTS

Our special thanks to the students involved in this study and the Faculty of Architecture and Urbanism of the Federal University of Juiz de Fora (FAU-UFJF) for financial support for presentation of this work. We would also like to thank the Coordination for the Improvement of Higher Education Personnel (CAPES), for granting scholarship abroad (PDSE 2016, Process nº 88881.132624/2016-01; FA/Ulisboa), complementing the development of this study.

REFERENCES

Arnheim, R. 2012. Arte e percepção visual: uma psicologia da visão criadora. São Paulo: Cengage Learning.
Lopes, R.F. 2017. "Sentir através de": o ensino do desenho de observação na Arquitetura e Urbanismo à luz da Fenomenologia da Percepção. PhD Thesis. Niterói: Federal University Fluminense.
Maturana, H.R. and Varela, F.J. 2001. A Árvore do Conhecimento: as bases biológicas da Compreensão Humana. São Paulo: Palas Athena.
Merleau-Ponty, M. 1971. Fenomenologia da Percepção. São Paulo: Freitas Bastos.
Mueller, C. 1977. Psicologia Sensorial. 2nd edn. Rio de Janeiro: Zahar Editores.
Ostrower, F. 2008. Criatividade e processos de criação. 23rd edn. Petrópolis: Vozes.
Pallasmaa, J. 2011. Os olhos da pele: a arquitetura e os sentidos. Porto Alegre: Bookman.
Pallasmaa, J. 2013. A imagem corporificada: imaginação e imaginário na arquitetura. Porto Alegre: Bookman.
Piaget, J. and Inhelder, B. 1993. A representação do espaço na criança. Porto Alegre: Artes Médicas.
Santaella, L. 2012. Percepção: fenomenologia, ecologia, semiótica. São Paulo: Cengage Learning.

Drawing (...) City (...) Body, Dwelling on Earth – Janeiro (Ed.)
© 2019 Taylor & Francis Group, London, ISBN 978-0-367-02602-8

Scarpa in Sicily: The memory of our senses

C. Giunta
Dipartimento di Architettura, Scuola Politecnica, Università degli Studi di Palermo, Palermo, Italy

ABSTRACT: For an architect thinking about his own city, and a specific piace inside it, to improve it means concentrating on the specific qualities of the site from a historical,economic, social and cultural standpoint. This writing will go into the details of this reflection, through the experiences of Carlo Scarpa and the prominent role undertaken by him in the definition of a possible innovative strategy dealing with the 1953 Messina Exhibition on Antonello. All this is directed towards the acquisition of the capacity for recognizing the physicality of a meaningful architecture and the spatial relationships that define it, like for example those between interior and exterior, thus making one's own the concept of threshold and spatial continuum. Design choices must be directed towards a process which has as its final objective the defining of a particular sequence linked to the idea of contemporary interiors.

Keywords: Contemporary interiors, spatial continuum, relationships

1 INTRODUCTION

The first work carried out by Carlo Scarpa in the land of Sicily was the design of the exhibition on Antonello, in 1953[1], which took place thanks to a fortunate meeting with one of the organizers, Roberto Calandra. In November 1952, in a pause during the INU Congress, Calandra went to the Correr Museum to see the exhibition on Toulouse-Lautrec, and was surprised by the simple oblique inner walls made of material hanging from the ceiling, which hid the structure of the building[2].

Roberto Calandra decided to contact the designer, not knowing that he was one of the most esteemed Venetian architects who, in pre-war years, had been appointed to design some exhibitions for the Biennale di Venezia[3]. So Calandra proposed and Scarpa accepted to go to Messina to design and set up the exhibition "Antonello and Fifteenth Century Painting". This was beginning of a sincere friendship which, after the exhibition organized in the city of the Strait, and the design and layout of Palazzo Abatellis, saw them involved in 1972 in a fraternal and professional association, for the restoration of Palazzo Chiaramonte, the so-called Steri (from Hosterium, fortified palace): this was another significant work, in the interior of one of the most representative monuments of the Sicilian capital[4].

1. Francesco Dal Co and Giuseppe Mazzariol, *Carlo Scarpa. Opera completa*, Electa, Milano 1984; see also Sandro Giordano's chart at p. 114; Sergio Polano, *Carlo Scarpa: Palazzo Abatellis. La galleria della Sicilia, Palermo 1953–54*, Electa, Milano 1989.
2. Francesco Dal Co, Introduction, in Sergio Polano, op. cit., p. 9. Regarding the figure of Roberto Calandra, see the degree thesis of Walter Leonardi, Carlo Scarpa and Roberto Calandra: Interventi siciliani tra progetto e restauro, relatore Sandra Paletto with Michela Benente, Politecnico di Torino, 2. Faculty of Architettura, Corso di laurea magistrale in architettura per la sostenibilità, 2012.
3. Orietta Lanzarini, Carlo Scarpa. L'architetto e le arti. Gli anni della Biennale di Venezia 1948–1972, Regione del Veneto- Marsilio, Venezia 2003, p. 76.
4. 1972–1998 The design of the Cattedra di Restauro of the University of Palermo directed by Roberto Calandra with Camillo Filangeri and Nino Vicari and consulenza di Carlo Scarpa. Cfr. Antonietta Iolanda Lima (edited by), Lo Steri di Palermo nel secondo Novecento. Dagli studi di Giuseppe Spatriano al progetto di Roberto Calandra con la consulenza di Carlo Scarpa, Diario Flaccovio Editore, Palermo 2006, pp. 11 ff.

Figure 1. Carlo Scarpa, the 1953 Messina Exhibition on Antonello.

The exhibition of Antonello's masterpieces, fitted out at palazzo Zanca, was widely heralded by critics and public[5]; as Lionello Venturi wrote, in Sicily everyone should be grateful, "especially to Fiocco, to Vigni and to Carandante"[6]. Scarpa's layout for the exhibition arranged many of Antonello's masterpieces in separate strips of pleated fabric.

We are dealing with a deliberately abstract and candid place, its purely spatial function denoted by the organization of the exhibition environment with slanting curtains. Scarpa used a light fabric, detached from the surface of the walls, to redefine the proportions of the space itself, so modifying the original character of the rooms of the building.

"The space generated appears alive and throbbing, coherent with the character of the works on display. The pre-existent walls are fully lined, the result of an extremely precise work. Scarpa constructed with his pleated veils a truncated pyramid, inside which are arranged the stands for the works of art, according to the effects of the Mediterranean light, which passing through the silk curtains, breaks onto the oblique walls. This theme, of an objective and

5. Gioacchino Barbera, "Carlo Scarpa a Messina: l'allestimento della mostra antonelliano (1953) e il progetto non realizzato del Museo Nazionale (1974–1976)", in Kurt W. Forster, Paola Marini, Studi su Carlo Scarpa, 2000–2002, Regione del Veneto-Marsilio, Venezia 2004, p. 289; see also Paola Marini, "Mostra Antonello da Messina e la pittura del '400 in Sicilia", in Guido Beltramini, Kurt W. Forster, Paola Marini (edited by), Carlo Scarpa. Mostre e musei 1944–1976. Case e paesaggi 1972–1978, Electa, Milano 2000, pp. 120–125. Matteo Iannello, Antonello da Messina e la pittura del '400 in Sicilia nell'allestimento di Carlo Scarpa e Roberto Calandra, in "Lexicon", Edizioni Caracol, n. 19, 2014, pp. 55–64; Santo Giunta, Carlo Scarpa. Una [curiosa] lama di luce, un gonfalone d'oro, le mani e un viso di donna. Riflessioni sul processo progettuale per l'allestimento di Palazzo Abatellis, 1953–1954, Marsilio, Venezia 2016.

6. Lionello Venturi, 1953, pp. 259–261, in Giorgio Vigni e Giovanni Carandente (edited by), Antonello da Messina e la pittura del '400 in Sicilia, Catalogo della mostra, Messina, Palazzo Comunale, 30 March-30 June 1953 published in the "Antologia di scritti su Antonello" (a cura di Riccardo Pacciani), Eugenio Battisti, Antonello. Il teatro sacro, gli spazi, la donna, Novecento, Palermo 1985, p. 293. Palazzo Collicola. Located in the historocal center fo Spoleto the same level also contains the Giovanni Carandente library, a national resources that houses thousands of books and art catalogues. Cfr. Gianluca Marziani, Palazzo Collisola. Visual arts, "Interni", n. 1/2 January-February 2018, pp. 82–83.

Figure 2. Antonello da Messina, The Sibiu Crucifixion.

vibrant lightness, has two Venetian precedents"[7]. The first instance referred to was that of the huge veils of orthogonal weave for the ceilings which defined the layout for the exhibition of the Rassegna d'Arte Contemporanea, in the Napoleonic wing of the Correr. The second was the layout already mentioned for the 1952 Toulouse-Lautrec exhibition[8].

Carlo Scarpa in these years, often resorted to visual mechanisms; they were real operations of composition which presented the itinerary of the exhibition in contemporary terms. At Messina too, the original form of the high rooms of the palazzo disappeared before the visitor's eyes, having been absorbed inside a truncated pyramidal space. A subtle light received Antonello's masterpieces on display, and in particular, as Giuseppe Fiocco wrote in his introduction to the catalogue: "We should be even more grateful in this regard to the generosity of Rumania, for having sent to the Exhibition, almost as compensation for the missing works, the precious picture from the Sibiu museum in Transylvania, of the Crucifixion, dramatically represented against the native background of the port of Messina"[9].

---

7. Among the works to be fitted out by Scarpa was the Pietà of the Correr Museum of which Bernard Berenson speaks as being, "a ray of light [which] illuminates the Pietà". See Bernard Berenson, 1955, pp. 9–12, published in the "Antologia di scritti su Antonello" (a cura di Riccardo Pacciani), Eugenio Battisti, op. cit., pp. 292–293.
Scarpa's layout for the exhibition arranged many of Antonello's masterpieces in separate strips of pleated fabric. "Una strizzatina d'occhio consapevole e decisa a un'opera dell'architetto Josef Hoffmann—allora commissario del padiglione austriaco—da lui realizzata nel 1934". Cfr. Philippe Duboÿ, Carlo Scarpa, L'art d'exposer, Association des Amis de la maison rouge, La maison rouge et jrp, Ringier Kunstverlag AG 2014. Cfr., trad. it., Carlo Scarpa l'arte di esporre, traduttore R. Rizzo, Johan & Levi Editore, Monza 2016, p. 107.
8. "Here appeared, perhaps for the first time the "weft and woof" motif of the ceiling velarium used subsequently for the layout of the exhibition in [....] Sicily". See Orietta Lanzarini, op. cit., p. 76.
9. From the Introduction by Giuseppe Fiocco, Catalogo della mostra, 1953, pp. 11–13, and published in the "Antologia di scritti su Antonello" (di Riccardo Pacciani), Eugenio Battisti, op. cit., pp. 290–292.

It is precisely this little Sibiu picture (39 cm.x23 cm.) that Scarpa used as a spatial hinge, which according to the exhibition structure, was arranged in a barycentric and slanting position with respect to one of the walls. Regarding this, Giovanni Carandente reminds us of some expedients practiced in the design of the Messina exhibition: "[...] An opening was made in the curtain of the corresponding window so that the visitor could compare the incredible painted microcosm with real life: the crescent-shaped port of Messina, which serves as a background to the Crucifixion scene with the actual crescent-shaped Zanclea, set against the same intense, vibrant blue of the sea, a glimpse of which can be had beyond the oblique wall covering"[10].

This seeking after the visual logic of the piece of artwork to be displayed, allows in this instance (only documented at present) for a re-invention of the space. This is re-formulated in relation to the works of art which, perceived in their hidden potential, compose the space. It is a visual expedient which connects the significance to the significant., making the view of the Crucifixion real. "He arranged the Pietà Correr—writes Carandente—on a stand a few centimetres high so that the perspective boldness could be better grasped (of Christ bending over as seen against the towering wings of the Angels) [...] When Bernard Berenson crossed the threshold of that exhibition room he was so moved that he reeled. He leant on my arm and I could feel him tremble. Roberto Longhi, who was less emotional than Berenson, wrote in his Frammento Siciliano that the design of the exhibition was "elegantly accomplished, as regards the scenic aspect, by the extremely skillful Scarpa""[11]. The great success and the refinement of Scarpa's design would lead Giorgio Vigni[12], curator together with Giovanni Carandente[13] of the Messina exhibition, to grant him the commission[14] of organizing a new museum layout for Palazzo Abatellis.

Carlo Scarpa arrived for the first time in Palermo in the Fifties[15]. In the city only a few automobiles circulated and the historic centre was the workshop of able craftsmen. There are no clear testimonies, or direct recollections regarding Scarpa's stay in the Sicilian capital, but he would certainly have seen the huge ficus macrophylla[16] in villa Garibaldi[17], a short distance from via Alloro.

He was aware of the new challenge which he was about to undertake in Sicily. In the space of a few months, (the museum would be inaugurated on the 23rd June 1954), he carried out a design strategy which, besides foreseeing the completion of the work, studied the system of routes for the exhibition and chose, with the help of Giorgio Vigni, the works to be displayed. As appears from the epistolary exchanges, Scarpa verified the spatial organization of Palazzo Abatellis while still in Venice.

---

10. Giovanni Carandente, "Vent'anni di lavoro", in Francesco dal Co, Giuseppe Mazzariol, op. cit., p. 202.
11. Ibidem.
12. Regarding the well-known epistolary collection between Giorgio Vigni and Carlo Scarpa see Paolo Morello, Palazzo Abatellis, Il maragma del Maestro Portulano da Matteo Carnilivari a Carlo Scarpa, Grafiche Vianello, Ponzano/Treviso 1989; in particular the documented analysis in the essay by the same author, "Il museo di Carlo Scarpa. Dal restauro del palazzo all'allestimento della Galleria", pp. 55.62. See also "Antologia epistolare", in Sergio Polano, Carlo Scarpa: Palazzo Abatellis. La galleria della Sicilia, op. cit., pp. 82–85.
13. Giovanni Carandente, "Vent'anni di lavoro", in Francesco Dal Co e Giuseppe Mazzariol, op. cit., pp. 202–206
14. "It was his habit to accept unremunerated work; but on that occasion he received an award of three hundred thousand lire, an extraordinary remuneration due to the great success of the exhibition and the refinement of his contribution". Paolo Morello, op. cit., p. 55.
15. Regarding the specific characteristics of a "Sicilian architecture" and in particular of architecture from Palermo of the twentieth century, see the Introduction of Vittorio Gregotti in Matteo Iannello, Blenda Scolaro, Palermo. Guida all'architettura del '900, Ed. Salvare Palermo, Palermo 2009, pp. 9–10.
16. The largest tree in Europe, 30 metres high with a circumference at the base of the trunk which measures more than 21 metres, its foliage having a diameter of 50 metres. See www.florablog.it/2009/12/15/alberi-monumentali-i-ficus-macrophylla-di-piazza-marina-palermo.
17. The design for the garden of Villa Garibaldi in piazza Marina was put into effect between 1861 and 1864 by Giovan Battista Basile.

From a study of his letters we note the self-assurance with which he dispensed his advice on the building work. His only pressing preoccupation was to know the objects to be displayed.

On the 27th October 1953 he wrote to the curator: "[...] when I'm in Palermo it's essential that you have a tangible certainty with regard to the placements, because we will have to study and choose the colours, fittings and mountings; it is not an exhibition that can be improvised, every single thing must be thought through and carried out in the best way possible so that the exhibition can last a certain length of time, while not being immutable, rather, on the contrary. The walls opposite the windows are always negative for exhibiting, so it is essential to provide material (wooden boards, planks, murals, half-murals) which will help us to put into place checks to find out where and how certain works of art should be placed, at what height and with what degree of inclination etc. etc."[18]; later on in the same letter, he repeats that "we are not dealing with a temporary exhibition"[19].

From the correspondence emerges a difficulty connected with the distance existing between Venice and Palermo. Vigni, urged on by the Ministry, in turn put pressure on Scarpa who felt this urgency to be an obstacle to the need for more time to carry out the preparations for the works to be displayed. It is a design, with a planning hermeneutics, which verifies the potential of the original structure of the palazzo and introduces some "fragments" which, as we shall see hereafter, reveal his experience as a designer of art exhibitions who is capable of transforming every object into "the apex of an imaginary perspective"[20].

With reference to this, Giovanni Carandente narrated a significant detail, which took place during the setting up of the Mondrian exhibition in 1956[21], held some years later at the Galleria Nazionale d'Arte Moderna di Roma: "I caught sight of him while he was trying, with the help of a custodian to lean against a wall the large painting of 1933, Losanga, with yellow square-shaped lines, but which should be seen obliquely on one of the corners. [...] After three failed attempts, he paused, satisfied. The painting finally tilted on the right edge, as confirmed by the signature and the date of the painter [...]. As Scarpa commented: "In other positions the painting lacked balance""[22]. Carandente became aware that "when he had to design the solution for the arrangement of a work of art, Scarpa loved, wherever possible, to handle the object or the painting, as though considering and examining the tactile sensation and the physical contact with the work, would predispose in him an ingenious result"[23].

This handling and investigating, therefore, in Scarpa's method was a normal procedure. Another witness of the period, Enrico Paolucci[24], commenting on the "sober elegance" of the rooms intended for the Metaphysical paintings in Venice in 1948[25], noted that: "Scarpa interrogated the work of art, always seeking to understand its internal rhythm, while never violating its spirit"[26]. Regarding this exhibition, a recent study by Orietta Lanzarini opens

18. Cfr. "Antologia epistolare", in Sergio Polano, Carlo Scarpa: Palazzo Abatellis. La galleria della Sicilia, op. cit., pp. 82–85.

19. Ibidem.

20. Giulio Carlo Argan, "Problemi di museografia", "Casabella continuità", n. 207, 1955, pp. 64–68, quoted in Marco Mulazzani, "Lo spazio interno come spazio architettato", in Antonella Hubel, Il museo italiano. The Italian Museum, with Prefazione by Sergio Polano, Edizioni Lybra Immagine, Milano 1997, pp. 59–75.

21. Design for the Exhibition Piet Mondriana, Galleria nazionale d'arte moderna di Valle Giulia, Roma, 1956. See the chart by Sandro Giordano in Francesco Dal Co e Giuseppe Mazzariol, op. cit., p. 119.

22. Giovanni Carandente, "Vent'anni di lavoro", Ivi, p. 204.

23. Ibidem.

24. Componente of Comitato Internazionale di Esperti della Biennale di Venezia dal 3 dicembre 1951, see note n.2 in Orietta Lanzarini, op. cit., p. 132.

25. The setting up of the XXIV Biennale, in the giardini di Castello, Venezia, 1948. Sandro Giordano writes in his chart: "on this occasion he is busy in the arrangement of seven rooms for the central pavilion; in particular, in the setting up of the exhibition "Tre pittori italiani dal 1910 al 1920" dedicated to Carrà, Morandi, de Chirico (halls III and IV), where the vertical walls covered with rich drapes contrast with the surfaces of oblique curtains" in Francesco Dal Co e Giuseppe Mazzariol, op. cit., p. 117.

26. Sandro Giordano, Il mestiere di Carlo Scarpa. Collaboratori, artigiani, committenti, tesi di Laurea A.A. 11983–84, relatore F. Dal Co, G. Mazzariol, IUAV, Venezia p. 111, quoted in Orietta Lanzarini, op. cit., p. 27.

up interesting points for future research and points out, with the help of some photos of the period, how "some objects recur from one picture to another, renewing the sensation, while following the itinerary of the exhibition, of a continuous déjà vu"[27].

This play of echoes in Scarpa's design process, which is not just figurative, is also present in the layout of the collections of Palazzo Abatellis, and represents a recurrent element in his poetics; a recurrence and a recognition which can by synthesized in a phrase used by Scarpa: "To obtain something, relationships must be invented"[28].

Profundity of thought and passion for the design of details represented for him the way to understand, prefigure and verify the nature of every solution and his trustworthiness represented a way of working based on an extraordinary sensitivity towards craftsmen's techniques, applied to the restoration of buildings and to their interior environments[29].

In particular, designing spatial devices for the eye of the visitor inside this historical building, indicates a sensitiveness regarding its places which is neither acritical nor casual. These are devices with a high aesthetic, cognitive and historical content, which harmoniously and with limpid simplicity interpret what exists and link the past and the present, with verifiable relationships confirming that the Abatellis is a timeless work of architecture.

Scarpa from the beginning was a man of his times, well-informed about the issues and experimentations in architecture and the figurative arts in the period between the two world wars, as is well-demonstrated by the recognizable influences in his first designs[30].

---

27. The yellow square of The engineer's lover, reappears in a still life, the bottle of this canvas has its reflection in that of Il gentiluomo briaco, while the oval form of the head of this character—with its full and melancholy profile—has its corresponding note in the profile of the sculpture, Maternità by Martini. Ivi. p. 31.

28. From the transcription of the Prolusione held on the occasion of the inauguration of the academic year 1964–1965 at the Istituto Universitario di Architecttura di Venezia and published with the title "Arredare", in Francesco Dal Co and Giuseppe Mazzariol, op. cit., p. 282.

29. After the years of his formation, which took place within the studio of the master glassmakers of Murano, where he created extraordinary objects of polychrome crystal of fabled transparency (Cfr. Marino Barovier, Carlo Scarpa. Venini 1932–1947, Skira, Milano 2015) he approached the design and layout of exhibitions and teaching through the fruitful contact with the Venetian Faculty of Giuseppe Samonà (Cfr. Ada Francesca Marcianò (edited by), Carlo Scarpa, Serie di Architettura/15, Zanichelli, Milano 1984, p. 8.). The period immediately prior to his Sicilian commitments was packed with assignments and work carried out: the restoration and arrangement of Ca' Foscari 1935–1937, the first public assignment of importance for thirty year old Scarpa; the setting up of the stands for the Venini society at the VII Triennale di Milano, 1940; the setting up of the exhibition "Arturo Martini", in cooperation with Mario Deluigi, the central pavilion of the Biennale, the gardens of Castello, Venice, 1942; the organization of the Gallerie dell'Accademia, Venezia,1945, other works for the Accademia in 1950 and in 1952, the arrangement of some canvasses in 1959; the setting up of the XXIV Biennale, the gardens of Castello, Venezia 1948, on the same occasion Scarpa set up the retrospective exhibition "Paul Klee 1879–1940" and with this exhibition began a long series of commitments in cooperation with the Biennial di Venezia (Cfr. Orietta Lanzarini, Carlo Scarpa. L'architetto e le arti. Gli anni della Biennale di Venezia 1948–1972, Regione del Veneto—Marsilio, Venezia 2003); the setting up of the new seat of La galleria d'arte moderna Il Cavallino, piscina di Arazzeria, Venezia, 1949; the setting up of the exhibition "Rassegna d'arte contemporanea", Napoleonic wing, Venezia, 1949; Pavilion of the book, giardino di Castello, Venezia, 1950; the setting up of the XXV Biennale, giardini di Castello, Venezia 1950; the setting up of the exhibition "I manifesti del Biennale", XXV Biennale, Napoleonic wing, Venezia 1950, Scarpa for this exhibition re-used as an exhibition structure the same one designed by him for the "Rassegna d'arte contemporanea"; the setting up the XXVI Biennale, giardini di Castello, Venezia, 1952; the setting up of the exhibition "Toulouse-Lautrec, Napoleonic wing, Venezia, 1952; the setting up of the exhibition of the INU, Ca' Giustiniani, Venezia, 1952; Project for Villa Zoppas, Conegliano (Treviso), 1953; Arrangement of the historical sections of the Correr Museum, piazza San Marco, Venezia, 1953. Straight after his work in Palermo, remaining in this same field, the assignments for some halls of the Gallerie degli Uffizi (1954–1956, with Ignazio Gardella and Giovanni Michelucci), for the Venezuela pavilion (1954–1956), for the enlargement of the Plaster cast exhibition Canoviana (1955–1957), for the Roman exhibition of Piet Mondrian (1956) and for the long, decennial work at Castelvecchio. See the catalogue of the works edited by Sergio Polano and the relative charts in Francesco Dal Co and Giuseppe Mazzariol, op. cit., pp. 97–114.

30. Marco De Michelis, "Ed anch'io son pittore". Ivi, p. 13.

He arrived in Palermo with a fund of knowledge connected to design, the experience of recent years and also to his being a man of craftsmanship[31] – we can think of his collaboration with the firm Vetri Soffiati Muranesi Venini & Co.[32], or his ability in the re-use of prepared structures (an example of which is the Rassegna d'Arte Contemporanea) – all opportunities which "return" in subsequent design layouts: from the use of slanting walls covered by a gathered fabric, to be seen in the hall of the Metafisici at the Biennale di Venezia[33] in 1948, and taken up again in other design layouts (as we have already mentioned for the exhibition on Antonello da Messina) – to his craftsman-like abilities in the use of wooden and iron stands, as in the layout for the sculpture La donna senza testa by Arturo Martini[34].

Carlo Scarpa was interested in the conceptual work of artists he met along his path, like Lucio Fontana, whom he met in 1948 at the XXIV Biennale di Venezia, on the occasion of the exhibition of his La scultura spaziale (1947), in black-coloured plaster[35].

To complete the picture, it is clear that mixing with other artists, for example the painter Mario Deluigi, whom he met during a course of studies at the Reale Accademia di Belle Arti di Venezia[36], gradually formed Scarpa's thought in a direction which was to become a decisive element. A comparison could be proposed, which would be interesting for certain analogies, between the solution for the layout of the Croce del maestro di Castelfiorentino and the theme of Spatialism. The position of the latter is a sort of "anticipatory" work made more evident in Concetti spaziali (1958–1968) by Fontana, where the cut on the canvas is a non-signifying sign which is recognizable in its three-dimensionality, as though the picture had lost its material limits.

On the other hand, Fontana's spatiality includes another dimension connected with temporality which passes through the cut and into infinity. The cut enables us to immediately understand the material. The image, and thus its dimension, is not defined by a formal scheme, but is a composition which fashions the material itself.

Placing once more our attention upon the visitor and his eye, within the route envisaged by Scarpa, if on the one hand he is attracted by past works of art, on the other hand he can understand the formal matrix of the environment in which he finds himself, reproduced in contemporary experiences. This calls to mind the considerations of Richard Sennett regarding "architecture as a work of art"[37], and of the awareness of the one who sees, who converses and holds a dialectic comparison between the exhibit and the space contemporary to us.

It is in this identification and comprehension between the container (the building) and the contents (the material exhibited) that Scarpa expresses his own original message by means of a modern lexicon[38].

The latter concept is essential in all the work of the Venetian maestro and, this can be understood in Palazzo Abatellis, as well as in the museum of Castelvecchio, where the visi-

31. A reference to a recognized social condition of our time, developed with attention in Richard Sennet, The craftsman, Yale University Press, New Haven & London 2008, (trad. it., L'uomo artigiano, Feltrinelli, Milano 2009).

32. Scarpa was artistic director in 1934.

33. It should be remembered that Scarpa began his collaboration with the XVIII Biennale di Venezia from 1932 and this lasted for 30 years. Orietta Lanzarini, op. cit., p. 23.

34. Arturo Martini e "Tre pittori metafisici dal 1910 al 1920", XXIV Biennale 1948. Cfr. Philippe Duboÿ, Carlo Scarpa, L'art d'exposer, Association des Amis de la maison rouge, La maison rouge et jrp, Ringier Kunstverlag AG 2014. Cfr., trad. it., Carlo Scarpa l'arte di esporre, traduttore R. Rizzo, Johan & Levi Editore, Monza 2016, p. 104.

35. Enrico Crispolti (edited by), Fontana, Catalogo generale, Electa, Milano 1986, p. 17.

36. On the relationships between Deluigi and Scarpa see Giovanni Bianchi, L'attività didattica di Mario Deluigi all'Istituto Universitario di Architettura di Venezia, in "Venezia Arti", 1966, pp. 95–106. See also by the same author Giovanni Bianchi, L'attività didattica di Mario Deluigi. Imparare a vedere l'arte, Edizione del Cavallino, Venezia 2010.

37. I refer to the shrewd comments published in Richard Sennet, The conscience of the eye. The design and social. Life of cities, Alfred A. Knopf, New York 1990, (trad. it., La coscienza dell'occhio, con uno scritto di copertina di Tomás Maldonado, Feltrinelli, Milano 1992).

38. Ada Francesca Marcianò (edited by), op cit., p. 48.

Figure 3. Il Gonfalone d'oro.

tor is dominated by the iconic vertiginous position of Cangrande, which however, in the interior becomes the figure of an empathetic background. In the Abatellis this empathetic communication from the inside has its beginning, as we have hinted, in the relationship which is created between the Croce del maestro del Castelfiorentino and the visitor, in a biunique relationship between the intrinsic figurative value of the work of art displayed and the one who is present at this "representation," by walking around it.

For the sake of completeness, the activity of the Spatialists should be stressed, which was made concrete in a series of programmatic manifestos[39], one of which would be presented at the XXIX Biennal di Venezia of 1958[40].

It is useful to make these points to enable the reader to approach the "spatial" aims of Scarpa and to some themes dealt with by Fontana[41]: "Art is a creation of man, it is not a physical act like eating or sleeping, it is a creation that can also finish at a certain moment or can be surpassed by other sciences, Art is none other than the evolution of thought, and if man's thought evolves to the point of becoming simplistic, then art is finished"[42].

39. The first manifesto was written between the end of 1947 and the beginning of 1948; the second was compiled on the 18th March 1948. Following these two texts there were: Movimento spaziale (2nd April 1950); Manifesto tecnico, Noi continuiamo l'evoluzione nell'arte, read at the IX Triennale of 1951; Manifesto dell'arte spaziale (26th November 1951); Manifesto del Movimento spaziale per la TV (17th May 1952); Manifesto spaziale del 1953, among the signatories, besides Fontana, were Deluigi, Morucchio and Guidi who were Scarpa's friends in the Thirties.

40. Another manifesto would be presented at the XXIX Biennale di Venezia in 1958.

41. Lucio Fontana on the occasion of the V Triennale di Milano (1933), collaborated with Luigi Figini and Gino Pollini in decorating the Villa studio for an artist, by carrying out La Bagnante nera and the plaster sculpture of Gli Amanti per la Casa del sabato per gli sposi, designed by the group BBPR (G.L. Banfi, L.B. Belgioioso, E. Peressutti, E.N. Rogers) and by Pietro Portalupi. In 1934 he again held a personal exhibition at Milione. In 1939 he participated with Franco Albini, Ignazio Gardella, Giulio Minoletti, Giancarlo Palanti and Giovanni Romano in the design for Il palazzo della Luce e dell'Elettricità per l'E42 in Rome (but never carried into effect), and other collaborations which we cannot refer to here because of limitations of space. In 1949 there was the first manifesto on spatialism, on the changes in the essence and in the form of art, the obsolescence of painting, sculpture and of poetry. Shortly after this, Fontana effected his first holes in the canvas. It is a step away from his "cuts" in the canvas in 1957.

42. Cfr. conversation with Fontana by Tommaso Trini, in "Domus", n.466, September 1968.

Figure 4. L'Annunziata di Antonello di Messina e il Gonfalone d'oro.

Just as though the reader were a detective who gathers clues, discovers examples and stories capable of corroborating our initial conjecture, the intention of this book is to propose an alternative hypothesis, listing point by point the assembled evidence in a convincing demonstration regarding the starting point of the exhibition route and so answer in a plausible way the preliminary question, "From where do you enter?" The conjecture, now more explicit because more-defended, is that you enter by going up to the first floor, through the hall of the Pisan Cross, whose sought-after position is the starting point of complex events[43].

If it is true that "the layout of Palazzo Abatellis in Palermo, carried out in 1953–1954, shows more than any in other work his reflections about space examined by metaphysics", as Orietta Lanzarini recently pointed out, it is also certain that we know little of the figurative references considered by the Venetian maestro[44].

Scarpa, having demolished the false ceiling made of wood, left the reinforced concrete girders visible, had the plaster stripped off the long walls of this rectangular room and on the walls' now visible stones, displayed some works of art[45].

Without iconographic infatuations, it is possible to look at the quality of the objects displayed in the spacious "hall of the Crosses—as Vincenzo Abbate reminds us—which is carefully measured in its harmony, in the juxtaposition between the precious fourteenth century

43. "Every "place" is a complete story, a cosmogonic event arranged in order to understand the origin of "things" the aspect of the change brought about by confining things to a museum. Let's call it the "metamorphoses exhibited" in the chaos/cosmos between: truth/lie, beauty/kitsch, art/document, figure/background, image/didactics, container/contents, collectionism/everyday life, history/virtuality. A provoking bipolarism, with an incessant critical tension and creative energy", in Pasquale Culotta, "Nella dinamica del caos", in Maria Clara Ruggeri Tricoli, I Fantasmi e le cose. La messa in scena della storia nella comunicazione museale, Lybra Immagine, Milano 2000, pp. 7–8.

44. Orietta Lanzarini, op. cit., pp. 31–32.

45. It is useful to take up again the writing of Giovanni Carandente: "The solutions which he found were all particularly happy ones, from the typical render used for exteriors—slaked lime, which was dear to him—to the frames of the polylobed windows which allowed a clear reading of the architectural design; from the singular solutions given to the roof structure of the great hall (the original ceiling collapsed under the bombs) to the beautiful open staircase, of Palladian inspiration on the ground floor". in Giovanni Carandente, "Vent'anni di lavoro", in Francesco Dal Co, Giuseppe Mazzariol, op. cit., p. 204.

Figure 5. Palazzo Abatellis. Il vaso Malaga e, sullo sfondo, il busto di Laurana.

paintings on wood, placed on the white backdrop of paneled walls, and the frescoes of Risalaimi, hung with their decorative motifs like drapes or tapestries"[46].

Carlo Scarpa is aware of all this, but places a really luminous object along the route of the visit, aligned with the small passageway, ennobled by a stone portal. It is the golden Gonfalon exhibited on the other side of the spatial ambit of the room of the Crosses, which seems to constitute another "clue".

Our study, from here on, will try and establish a relationship, and not just a visual one, between the Great Hall of the fourteenth century, La Sibiu Crucifixion[47] and The Gonfalon.

In a way that is not too veiled, Scarpa entrusts a highly symbolical significance to this exhibit. It deals with a trail, a connection with that little painting which, in his design of the Antonello Exhibition of 1953, managed to enrich the relationship between the past and present; in fact, by means of an opening in the curtains fitted by him, the visitor perceived the microcosm painted in the fifteenth century against the real background of the marine landscape of Messina which is represented in it. A continuity with this painting is made conceptually recognizable through associations, contrasts and similitudes by means of an original use of its evocation in this exhibit, which appeals to its primary character as a devotional object, probably carried in procession as the standard of a confraternity.

The well-educated visitor, who here looks beyond the Crosses, makes out the wooden form of a Gonfalon, completely covered in pure gold. Inside it he can recognize the dimensions of the Sibiu painting in the small empty square, placed against the background of a green panel, and with an acquired knowledge, he calculates, meditates and, because of the presence (in the educated memory)/absence (from the exhibition) of the Sibiu painting, recognizes its proportions in the figure (The Sibiu Crucifixion is 39 cm high by 23.5cm wide), which in fact the Gonfalon would have been able to contain[48].

46. See the bibliography of Vincenzo Abate's essay, "Le collezioni della Galleria", in Sergio Polano, Carlo Scarpa: Palazzo Abatellis. La Galleria della Sicilia, op. cit., pp. 65–77.

47. Refer to the initial writing on the 1953 exhibition at Messina.

48. "Although the Crucifixions of Antonello—writes Eugenio Battisti—don't present evident novelties from the iconographic point of view, and seem to perform a simple devotional function, this function is immediately complicated by their reduced dimensions (that of Sibiu measures 39 cm by 23,5 cm), making them inappropriate for collective functions, which are susceptible to emotional and sentimental abandon. Instead they are more suitable—even if they had been requested as the standard for a confraternity —for meditation and silence, with the consequent effects of a meticulous and unhurried exploration of every detail". Cfr. Eugenio Battisti, op. cit., p. 17.

Figure 6. Uscita, fra il verde e l'azzurro, alla sala del Laurana.

Figure 7. Ultima sala.

It is a reference by analogy to what is represented in Antonello's painting, which identifies a perspective system, formed by two parallel low walls, similar to the unplastered walls of this room.

Scarpa achieves here a spatiality based on the depth and on the chromatic output of the long lateral walls, which compress the space in a similar way to the pictorial plan studied by Antonello, where Christ is made to appear in the foreground by placing the two crosses on either side of him[49].

49. For the description of the painting, see Maria Grazia Paolini, Antonello da Messina e la cultura fiamminga del primo Quattrocento, Istituto di Storia dell'Arte dell'Università di Palermo, Course lecture notes for the academic year 1968–69, p. 46 and published in the "Antologia di scritti su Antonello" (edited by Riccardo Pacciani), Ivi, p. 282.

Figure 8. La fonte di vita e l'uscita del percorso. A destra la panca dove sedersi per scorgere l'azzurro del cielo.

Keeping the back walls white is the result of an architectural construction which, according the colour of the unplastered walls, re-evokes a new landscape where the horizon line and the penetration of light are equivalent to that of the exhibited items.

Is it perhaps according to the same principle that the Crosses are placed on a parallel with the shorter wall and that the flooring is tiled with terracotta, in a colour similar to the ground in Sibiu's painting?

The viewer, looking at the Gonfalon finds himself inside a "transferred" temporal articulation, which recalls the memory of the sorrowful figures of the Antonello painting: the wide expanse enclosed between the horizon line and the gulf of Messina, with the view of the Peloritan mountains and the Eolian islands in the background, here becomes a conceptual line for the collocation of the art-works.

According to what we have just said, one could suppose that the Venetian maestro read into the essence of the lines of construction of the Sibiu Crucifixion[50] the harmonic elaboration linked to the figure of the Crosses. Is it a reference to the whole, in harmony with the theme of his layout, and to the "things" contained in the painting, in which the perspective theme gels and materializes on the horizon where, from close-up, a small detail is distinguishable: ships in roads with their sails lowered. Perhaps this is the answer to another clue?

It is the creation of a nebula, perhaps unconsciously, in the eyes of the visitor who doesn't know the picture, but who shortly would be able to discover it in detail, with, hopefully, audio-visual aids or simply through images in a catalogue.

Scarpa tried to insert an element, with some changes in scale that are functional not only to the strategy of the layout, which suggests the characteristic traits of the compositional procedure, This is another clue that is measurable in relationship to the figure (the Gonfalon) and the clearly visible background (the colour green) in the next room dedicated to Antonello.

---

50. Eugenio Battisti writes: "In the Crucifixion of Sibiu the splendid view, having precise topographical connotations, should have a reference to the clients, that is to a confraternity, or to private clients particularly interested in maritime navigation, because of the reference to the port of Messina, which still today, notwithstanding its renovation, maintains the same structure of the curve of a circle, with the lighthouse and the castle and where an armed ship in roads, is visible, ready to unfurl its sails", Ivi, p. 21.

The holder displayed is in pure gold and is a rare and precious object, which we know originated in Tusa (Messina). It recounts the compositional and formal tastes of the artist rather than the craftsman, who inspired these rich, ligneous and plastic decorations, typical of palaces and religious buildings. The style of this Gonfalon is the result of the delicate work of a wood-carver, capable of translating, in reduced form, the elegant models of classical monumental architecture with particular care. The visitor is compelled to concentrate his attention not only on the richness of the carving, but he must also know how to grasp the inner energy of the missing picture. The method to be adopted for displaying the gonfalon is indicated, beginning with the object itself, by the absence of the picture.

The paths of Antonello and Scarpa meet up, highlighting the correspondence that exists between the former's precision as a miniaturist and the latter's meticulous wishes to put into effect a perspective device within a spatial complexity.

With surprising effect, Scarpa arranges the Gonfalon as a framework of pictorial imitation. Seen from a distance, in fact, it becomes a reference element for a possible journey into three-dimensionality which anticipates the world of the artist which follows: Antonello da Messina.

The result, well-balanced between the quality of the materials used and the sought-after geometry, is the outcome of a continuous play of compressions and dilations which will also be found in the following space. I think it is superfluous to indicate that here "the Sibiu Crucifixion" are not the expected reference, but are a certain presence within the memory of our senses.

REFERENCES

Antonietta I.L. 2006. edited by, Lo Steri di Palermo nel secondo Novecento. Dagli studi di Giuseppe Spatriano al progetto di Roberto Calandra con la consulenza di Carlo Scarpa, Diario Flaccovio Editore, Palermo.

Beltramini, G., Forster, K.W., Marini, P. 2000. edited by, Carlo Scarpa. Mostre e musei 1944–1976. Case e paesaggi 1972–1978, Electa, Milano.

Dal Co, F. and Mazzariol, G.1984. Carlo Scarpa. Opera completa, Electa, Milano.

Duboÿ, P. 2016. Carlo Scarpa, L'art d'exposer, Association des Amis de la maison rouge, La maison rouge et jrp, Ringier Kunstverlag AG 2014. Cfr., trad. it., Carlo Scarpa l'arte di esporre, traduttore R. Rizzo, Johan & Levi Editore, Monza 2016.

Forster, K.W., Marini, P 2004. Studi su Carlo Scarpa, 2000–2002, Regione del Veneto-Marsilio, Venezia.

Giunta, S. 2016. Carlo Scarpa. Una [curiosa] lama di luce, un gonfalone d'oro, le mani e un viso di donna. Riflessioni sul processo progettuale per l'allestimento di Palazzo Abatellis, 1953–1954, Marsilio, Venezia 2016.

Iannello, M. 2014. Antonello da Messina e la pittura del '400 in Sicilia nell'allestimento di Carlo Scarpa e Roberto Calandra, in "Lexicon", Edizioni Caracol, n. 19.

Lanzarini, O. 2003., Carlo Scarpa. L'architetto e le arti. Gli anni della Biennale di Venezia 1948–1972, Regione del Veneto- Marsilio, Venezia 2003.

Marcianò, A.F. 1984. edited by, Carlo Scarpa, Serie di Architettura/15, Zanichelli, Milano.

Morello, P. 1989. Palazzo Abatellis, Il maragma del Maestro Portulano da Matteo Carnilivari a Carlo Scarpa, Grafiche Vianello, Ponzano/Treviso.

Polano, S. 1989. Carlo Scarpa: Palazzo Abatellis. La galleria della Sicilia, Palermo 1953–54, Electa, Milano.

Drawing (…) City (…) Body, Dwelling on Earth – Janeiro (Ed.)

Literary and travel images for the knowledge of the city

V. Cardone
Department of Civil Engineering, University of Salerno, Fisciano, Italy

S. D'Auria
Engineering Macroarea, University of Rome Tor Vergata, Rome, Italy

E. De Feo
Department of Civil, Environmental and Architectural Engineering, University of Padova, Padova, Italy

ABSTRACT: One of the most fruitful approaches to the study of the city is in the analysis of travel reports, both graphic and literary.

The journey to an unknown city, in fact, always involves a careful exploration—that cannot be achieved without both a physical and corporeal relationship, in which the foreign eyes manage to catch things that escape its inhabitants—and the need to remember, document, analyze, interpret, understand, represent and tell it, for oneself and for others.

In this sense, they are also fundamental the descriptions of cities in the literary works of some writers, that have linked their name to some particular city, in which they have placed scenarios of the narrated adventures, constructing or consolidating their myth or re-inventing it. In some cases they have represented it only in the topographic, architectural and spatial articulation; in other they have attempted to reveal its pulsating, underground, obscure life, to convey the unique and unrepeatable atmosphere that gives it life, coming to grasp the mysterious charm of his teeming humanity, the inexplicable imbalance or the incredible harmony between his body and his soul.

Such intense relationships have been established precisely because in many works of these authors the city is protagonist—as well as the characters, sometimes more than the characters but often thanks to their movement in its world universe and their relations with its geometrical space; as a subject and as an object: an organism supported by intricate spatial patterns, that assure its vital functions and allow its independent existence, and set of human events, which develop in it.

Keywords: cities, images, travel drawings, travel literature, literary images

1 INTRODUCTION

The city is an expression of human creativity, perhaps the most important product which stimulated and still now stimulates the interest of humankind. The city was firstly field of interest of philosophers and poets, historians and geographers, political reformers and utopians, before the claim of architects and planners of a competence almost exclusively over this matter.

Some scholars remember that the city *is* "already before its birth, an ambiguous and contradictory sign, [because] it originates from a violent end [and the metaphysics of the city] if investigates the origin, it comes across the myth of Cain, and a city as a legacy of a fault, the space in which the exile takes place" (Zarone, 1994: 79–80).

Almost everyone is of the opinion that the history of the world is an urban history, and that civilization represents the triumph of the city. Since the beginning of the last century, the conviction that the modern city is "the hallmark of modern man, his large format portrait" is firmly rooted. "Man built himself, the great modern city is the work of human imagination. The metropolis is manifestation of choices made by human beings" (Schabert, 1994: 17).

Urban geographers consider the city as a complex social organization, product of an advanced civilization. The complexity of the topic involves all fields: historical, social, cultural, economic, architectural, technological, etc.

For this reason the study of the city is therefore more than complex and cannot be reduced to the analysis of architectures, urban space, individual items or of functions and relationship. Specialists from various disciplines have taken up the subject also through multy-interdisciplinary studies, finding a perspective for the right approach and key points for the reading of the city, mindful that no studies provided a comprehensive knowledge.

One of the most fruitful approaches to the study of the city is the analysis of travel reports, not only graphic—as, first of all, the drawings of architects travelers—but also literary: a large quantity of material, mostly unexplored.

Travelers, as they are unfamiliar with the cities, are best able to capture aspects of the context that normally elude ordinary perception (Cardone, 2007; Id., 2017).

The perception of space in fact is influenced by the culture of those who experience it. It is a social fact (Nuvolati, 2013: 69). It has been noted for decades that the city "cannot be identified with buildings, whether centrally located or outlying, rather in the mobility, in the flow of cars, finance and information" (Duque, 1994: 52).

And before that, in human movement; the same perception is linked to mobility, because, "by walking we perceive the space, we measure and it internalize it" (Pavia, 2015: 5).

And still more:

> *"Human existence is movement and this movement creates spaces. To be more precise: human life consists of a thousand movements and these movements create the space in which they exist. Their existence is a spatial existence and so is their 'space'. It is the space in which they exist: human space. Yet this space, human space, is not a space like any other. It is a flowing space, that flows continuously like the human movements that create it. Human existence is a spatial existence and, as such, it is a social existence. [...] Its structures are anything but solid and its outlines are anything but clear. Whatever its forms may be, these are inevitably transformed" (Schabert, 1994: 23).*

These processes take place through the movement of the human body or of something commensurate with it, even when the man is in a static position or moves using mechanical devices: the car, for example, as the research group of Kevin Lynch proved more than half a century ago, in a study that has not enjoyed the good fortune it deserved (Appleyard, Lynch and Myers, 1964).

According to some scholars, the spaces of western cities, especially those with historical origin, can be traced back to two main iconic archetypes, to which it is possible to refer their multiple figurative declinations: the scene and the labyrinth (Cervellini, 2017).

Figure 1. Orhan Pamuk, Labyrinth.

The labyrinth is the inextricable tangle of the streets, devoid of a single center but characterized by multipolarity. To be known, this space can not only be looked at, but it must be penetrated, explored in different directions of its uncontrollable and unknown extension; choosing between alternative and contrasting options, retracing our steps, following twisted itineraries, with repeated feelings of disorientation.

From time to time, in this intricate tangle, a widening opens up allowing one to recover from disorientation and to find oneself: the square. Namely the en plain-air space, equipped with a main center; the scene, a manifestation of the controllable geometric order, pursued with the central perspective. It is the stage, in fact, for the celebration of the main rites, civil and religious, for the representation of existence and of the human condition.

It represents "the heritage of citizens, it is not enough just to cross it, we must take possession of it by remaining there. [...] But the cities of the Mediterranean area show that they need more than the presence, that is, a square without traffic and coffee" (Neumeister, 1994: 67).

In the square everything is apparently under control, observable and usable statically, yet we need to walk through it to take possession of it because its space is not static, but evolves. Both these archetypes have enjoyed great fortune among the scholars who have analyzed the city; but the scene, the square, can be considered part of the labyrinth, which is therefore the most suitable metaphor of the stratified city.

Walter Benjamin, also used the image of the labyrinth to describe modern cities, noting that "getting lost in it, in the same way as you get lost in a forest, it's something to learn".

Already the abbot Laugier considered the city as a forest and the condition of disorientation that the great cities generate, even in many of their inhabitants, has been the object of repeated reflection. A few years ago Claudio Magris wrote that "the forest of the modern traveller is the city, with its deserts and its oases, its voices and its solitude, its skyscrapers or its suburban taverns, its rectilinear streets which leads towards infinity" (Magris, 2005: xvii).

Other authors used the metaphor of the labyrinth, beginning with Jorge Luis Borges who used it in various writings and argued that the only way to know a city is by walking through it, by exploring it and by getting lost in it (Borges, 1952; Grau, 1997). But it is precisely Benjamin who provides the most effective and suggestive formulation, related in particular to the exploration of the labyrinth, and to the figure of the flâneur dear to Baudelaire to the practice of flânerie, which mainly characterizes the attitude of the cultured traveler.

For this reason, travel descriptions provide a series of 'portraits' indispensable for capturing the identity of a city.

The journey to an unknown city always generates the need to analyze it, to interpret, to understand, to document, to remember, to tell and to represent it, for oneself and for others. A city can be analyzed and described through images, literary and visual; the first in written form, the others consist of pencil or tempera drawings, or by image once obtained from a camera, while today the continuous stream of images, does not allow time for reflection: in this way images are used only for documentation, while the analysis is delegated to end users.

The drawings, in particular, often are the completion of the written récit; they provide feedback, clarifications and offer the possibility of ideally transporting people to the place described; they evoke feelings and arouse, in those who observe them, personal enjoyments, not mediated by the opinion of others. They also play an essential role in making the travel experience concrete otherwise relegated to a certain state of indeterminacy or left to the imagination of the reader.

The author Michel Chaillou observed that "dans les récits de voyages, même s'ils ne sont pas dus à des écrivains, il y a une aventure. [...] L'aventure que contiennent les récits de voyages procède de l'exploration. Un auteur de récits de voyages est un dévoreur d'espace, quelqu'un qui fait connaître des choses nouvelles de l'espace" (Chaillou, 1992: 79). It follows that the reader creates spaces from what he reads on the basis of imagination, because the story stimulates a spatio-temporal construction with few constraints and many degrees of freedom unlike a graphic image.

> "The eye also imprisons the gaze in a closed frame, while listening frees the spirit from too particular stimulations and increases freedom in relation to the environment

in this sense. [...] Logocentrism, that privileges speech over writing as a form of communication, opposes the indefinite creativity of poetic language to the relative immobility and poverty of visual representation, forced into its spatiality, and enhances multiple meanings of words against the apparently unique sense of images" (Wunenburger, 1999: 30).

The graphic image contextualizes the experience and limits it in a given space, reducing its bearing and, in some ways, arbitrariness. In this sense it assumes the role of control and a documentary proof of the experience. It is "a means of preserving and fixing the experience impressed into memory, to record the presence of the observer and the observed. [...] It connects past and present because it allows you to stop time, to freeze the moment, to stop the eternal flow of light" (Leed, 2001: 18–19).

The graphic image thus becomes a new information of external reality, turning into an even more stable information than what it represents.

> *"In fact, while reality continues to transform itself, the drawing stops this transformation and freezes it in a precise time and in a precise place. [...] The graphic sign can then be the only possible perceptive connection of an evolving culture with its origins and, in a changing world, only representation allows comparison with the past" (de Rubertis, 1994: 23).*

Writings and images are complementary, not equivalent communication channels. Cesare de Seta wrote that "we know very well that the rules of the figurative text and the literary text are very different. But between a verbal image and a visual image a syncretism is realized that takes shape in the consciousness of those who read a text or observe a painting. As in a play of mirrors, the two texts are compared along a common axis of balance" (de Seta 1999: 10).

Their joint use for the description of the city dates back to Albrecht Dürer. Between the fifteenth and sixteenth centuries, he visited almost all the fundamental places for the formation of an artist, leaving an immense amount of graphics (drawings, engravings, watercolors), but also written texts.

He sometimes used the written form and the drawing on the same sheet. Thus he anticipated a large group of travelers who later used both forms of expression.

Dürer's innovative contribution to the theme of travel drawing is precisely in the representation of the landscape and the city. He proposes, in fact,

> "a new cosmographic way of representing the landscape that reveals the action of the forces of nature through the observation of physiognomy, color and light of sites. The result is an image of timeless modernity. Moreover, for the first time the landscape acquired autonomous dignity, rather than being a frame for sacred or profane images" (Herrmann Fiore, 2007: 191).

Figure 2. Albrecht Dürer, houses on the Pegnitz at Nuremberg.

In the representation of the city, in particular, Dürer did not draw the urban center as isolated, common practice at the time, but he also created a faithful image by inserting it into the surrounding landscape. Subsequently, the graphic images have acquired a strong conditioning power in the thought of their users. A power comparable, or greater than the written texts as the experience of the Grand Tour travelers has shown. Many of them have made their images based on the belief that the drawing provides "a greater capacity to memorize the elements studied, becoming a useful tool for the study of the internal structure of things" (Pratelli, 1989: 100).

Thus the folder of drawing sheets and the box of watercolor colors become essential elements of travelers. Johann Wolfgang Goethe decided to learn how to draw before his Italian journey. The scholars who analyzed his travel sketches "an analytical instrument of knowledge of nature, of the landscape whether natural or human-built" (de Seta, 1982: 251).

It was not an isolated attitude or the habit of a great person. "The paintings allow us to identify the perceptible with what can be drawn. Painting and drawing were practiced as a means of training observation skills and stimulating visual intelligence" (Leed, 2001: 19).

A lot of detailed and in-depth research has been carried out on drawing travelers. These researches have produced serious studies, but sometimes limited in time and space, without a broad contextualization and a global vision. Some general publications on the topic of travel drawings are recent (Barba-Messina, 2005). Perhaps only for the architects' travel drawings a study was undertaken in this sense in the last quarter of the last century. however, even this activity is still partial (Cometa, 1999; Mangone, 2002; Moleón, 2003; Maglio, 2009; Cardone, 2014).

Even the most anonymous and amateur album of travel drawings constitutes, independently of artistic value, a precious document, more than many written stories, to study the evolution of the image of the visited cities. These drawings are personal notes and this characteristic, that distinguishes a large part of this production, gives it a more user-friendly look than the written page—result of long reflection, even with the help of guides and classic texts.

The short time lapse between the capture of a particular aspect of the place visited and its transfer on paper makes sketchbooks and the drawing of carnets de voyages less conditioned by the consolidated images of a city. For this reason, they reveal particular aspects that are invisible to the eye of the usual observer and of the artist who works for a long times.

However, both the writings and the graphics and the reciprocal contaminations must be studied.

> *"The drawer and the painter look and describe the reality transforming their analysis in synthetic images, the writer describes things through words and has the opportunity to narrate places and facts that cannot be represented. The brush and the pen are*

Figure 3. Federico García Lorca, Urban perspective with self-portrait.

different probes to give us the knowledge of a fantastic imaginary that makes real the moment painted or narrated. The same experience described and represented in different ways with two specular artistic forms generates an ambiguous attraction in the reader" (de Seta, 1999:10).

The last considerations open interesting scenarios for the study of travel drawings. The first, and the the most intriguing, is to examine the relationships between written text and drawings of the same traveler. We have more precious testimonies to analyze. One of the authors has already dealt with it in general (Cardone, 2006) and specific terms, focusing on the travel reports, writings and graphics of the French architect Pascal-Xavier Coste (Cardone, 2005), of Victor Hugo, of Benito Pérez Galdós, the Spanish novelist of Nineteenth century and author of splendid relatos de viajes (Cardone, 1993).

There are many artists who have used both expressive means during their travels: from the above-mentioned Goethe to Hans Christian Andersen, from Hermann Hesse to Evelyn Arthur Waugh, from the American Charles Dana Gibson—a satirist and illustrator of books, who made travel sketches, accompanied by a small text, a very personal form of expression, which had a great publishing success—to Czechoslovakian Karel Čapek (who coined the term 'robot'), author of travel stories accompanied by little sketches, to Tahar Ben Jelloun. Federico García Lorca should be also mentioned. He was not a great traveler—maybe he did not have the time, but surely the desire as his stays in New York, Cuba and Argentina suggest—nor he was the author of relatos de viajes.

He enjoyed drawing (Hernández 1998). His drawings of New York—the city where the traveler captures *"arquitectura extrahumana y ritmo furioso. Geometría y angustia" – remember the settings of Metropolis di Fritz Lang. They* are unavoidable from his experience and communicate much more than thousands of pages written on the metropolis of skyscrapers (Villanueva, 2008: 217–244).

The travel drawings of the writers have more value than the travel drawings of architects or any other traveler. They also served as a source of inspiration for subsequent writings, to

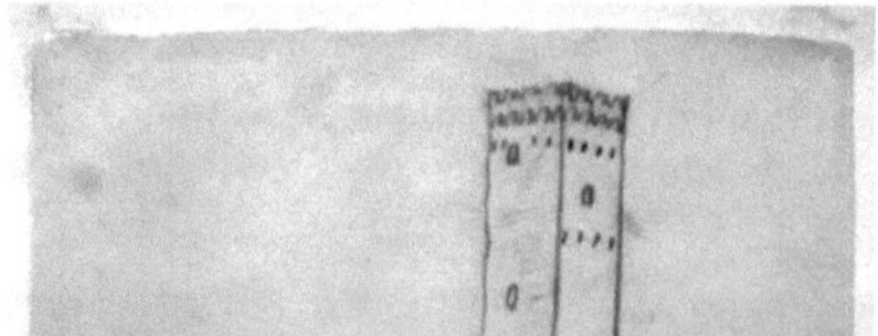

Figure 4. Hans Christian Andersen, leaning towers in Bologna.

create scenarios, locations, atmospheres, as well as to remember: it should be sufficient to recall Italian drawings of Andersen, who set in the Peninsula the novel de Improvvisatoren (Berni, 2013).

Therefore, these schematic drawings capture aspects that are not objective but related to the individual fruition, in a certain sense immaterial, that other travel drawings do not possess.

For the same reason, the descriptions of cities in the literary works of some writers are also fundamental. These authors have linked their name to some particular city, which they described, in which they placed the scenarios, sometimes still recognizable, of the narrated adventures.

They have thus succeeded in creating or consolidating the myth of a city reinventing it, whether when they represented it in its topographical, architectural and spatial articulation, or instead when they tried to reveal its pulsating and underground, obscure life, hidden from sight creating a unique and interesting atmosphere, reaching its soul.

Some cites have created writers, attracting them with their enchanting location, their splendid architectures, the strident contrasts between the regular geometries and the intricate system of their fabric, their past and their present, the dazzling lights and the enveloping shadows, the magical silence of their secret spaces, the deafening sound of their incessant breath, the mysterious charm of their teeming humanity, the inexplicable imbalance or the incredible harmony between their body and their soul.

In one case as in the other, writers and cities are bound in indissoluble relationships. For istance Fëdor Dostoevskij, Nikolaj Vasil'evič Gogol' or Aleksandr Sergeevič Puskin and Saint Petersburg; Charles Dickens and London; Victor Hugo, Honoré de Balzac, Charles Baudelaire, Émile Zola or Marcel Proust and Paris; Benito Pérez Galdós and Madrid; Franz Kafka and Prague; James Joyce and Dublin; Costantino Kafavis and Alexandria; Pier Paolo Pasolini and Rome; Jorge Borges or Ernesto Sábato and Buenos Aires; Ferdinando Pessoa and Lisbon; Anna Maria Ortese and Naples; Manuel Vásquez Montalbán and Barcelona; Jorge Amado and Salvador de Bahia; Don DeLillo or Paul Auster and New York, Orhan Pamuk and Istanbul.

Such intense relationships have been established precisely because the works of these authors have as their protagonist—alone or usually together with the characters and sometimes more than the characters—the city, as the subject and the object: an organism characterized by intricate spatial circuits that assure its vital functions and allow its independent existence. The city is the stage of human events. In some cases, the identity of a city, its collective imagination and perception still today coincide with those that the authors have built.

These descriptions are very important to know the city The artist, in fact, endowed with highly sensitive antennas, possesses a boundless imagination, determined also by cultural factors but not conditioned by physical and material limits related to the physicality and materiality of things, to their concrete technological and structural feasibility, etc.

Figure 5. Émile Zola, plan view.

Most of them have a strong connection to the body, because they are not descriptions of ideal or simply fantastic or absolute spaces—like those of Italo Calvino in Città invisibili, often mentioned inappropriately—but real spaces, perceived by man, in close relation with his movement. The story is almost always linked to man; there is always someone, the narrator or some character of the literary work, which describes the paths that follow, the views he makes, his personal fruition.

Many examples could be given, with the writers mentioned above. Particularly interesting is the case of Benito Pérez Galdós, whose graphic production has only recently been discovered.

Galdós, however, has always been drawing, with particular interest for architectural themes, in fact some of his sketches seem surveys of constructive details (Miller, 2001).

The same can be said for Victor Hugo, whose skill as a painter and drawer was already well known not only to his friends: although in 1863 the engravings of some of his drawings were published, with a presentation by Théophile Gautier (Hugo, 1863), his great graphic production—in which the theme of architecture is forcefully imposed together with beautiful landscapes, pen or colored, and imaginary images—was in fact only disclosed on the occasion of the centenary of death (1985) and of the bicentenary of the birth (2002) of the artist (AA.VV. 1985; Georgel, 2002). A study of his written reportages and travel images—of Spain, of the Rhine, of the Pyrenees, of the Belgium or France and of all the place he visited, leaving testimony—is started, partly extended to his entire and immense literary and graphic production.

Victor Hugo was a close observer of the choses vue, that he described with overwhelming charge, rereading and representing them in a real and fantastic way. The techniques used for his travel drawings, on which we can see corrections to the original drawing, testify to a broad reflection subsequent to their realization.

The best-known example of writer drawer is perhaps Émile Zola, the greatest exponent of naturalism, who went around the city to take notes, also graphical and photographic, for his novels and created thick Carnets (later published), thanks to which he was even considered an ethnographer.

He was convinced that the characters depend on the environment so he not only observed the reality he had to describe but also he experienced environments and situations. For the settings of his novels, he walks through Paris like a flâneur, through avenues, boulevards, markets, shopping malls and for those passages so dear to Benjamin, who wrote a large and unpublished metropolitan description, unfortunately unfinished.

Zola explores the places—the city and its architecture—meticulously, street by street, room by room, corner by corner, down to the smallest details, including installations and furnishings; he draws plans of streets, squares, urban areas, houses, facades, gates; he notes the functions that take place in a space, which will rework in his creative writing and that have nothing to envy to the final written text (Reim, 2001). He spends whole days in the places he wants to investigate and often these are the real protagonists of some of his novels: like the Les Halles market, in Le ventre de Paris, the third volume of Les Rougon-Macquart.

Today the most representative of these writers is undoubtedly Orhan Pamuk, who as a young man aspired to be an architect (like Paul Valery, another writer-drawer) and until the age of twenty-two, before opting for literature, he devoted himself to painting; he has nevertheless preserved a good figurative approach: to the point that many critics consider him a 'visual writer'. In his works on Instabul he gives new interpretation of the labyrinth after Benjamin and Borges.

We also like to recall the pages of Manuel Vásquez Montalbán on Barcelona, perhaps the best things about the cycle on the investigator Pepe Carvalho.

An in-depth analysis of these authors and their descriptions of cities, literary and graphic, however, is a complex subject which goes beyond these brief notes. They are only a general introduction to the theme. We will make it the subject of specific reports, on other occasions.

REFERENCES

AA.VV. (1985) Hugo. Dessins, Paris: Gallimard.

Appleyard, D. Lynch, K.& M., John R.. 1964. The Views from the Road. Cambridge: Mit Press.

Barba, S. Messina, B. (a cura di). 2005. Il disegno dei viaggiatori. Salerno: Cues.

Berni, B. 2013. I disegni italiani di Andersen, In Andersen Hans Christian, L'improvvisatore, trad. it. de Improvvisatoren. Roma: Elliot.

Borges, J.L. 1952. El Aleph, Buenos Aires: Losada.

Cardone, V. 1993. Introduzione, in Pérez Galdós Benito, Le città italiane. Napoli: Alfredo Guida Editore, pp. 5–23.

Cardone, V. 2005. Pascal Coste, viaggiatore d'architettura per lavoro, per studio, per passione, in Barba e Messina (a cura di) (2005), pp. 69–92.

Cardone, V. 2006. Disegno e scrittura, in Gambardella Carmine e Martusciello Sabina (a cura di), Le vie dei mercanti. Disegno come tipologia della mente, Atti del Terzo Forum "Le vie dei Mercanti", Capri 6–7–8 giugno 2005. Firenze: Alinea, volume "Relazioni", pp. 53–79.

Cardone, V. 2007. Le reti delle città nei viaggi della conoscenza, in Gambardella Carmine e Martusciello Sabina (a cura di), Le vie dei mercanti. Città rete_rete di città, Atti del Quarto Forum Internazionale di Studi. Napoli: La Scuola di Pitagora editrice, volume "Relazioni", pp. 47–64.

Cardone, V. 2014. Viaggiatori d'architettura in Italia. Da Brunelleschi a Charles Garnier. Padova: Libreriauniversitaria.it e Università degli Studi di Salerno.

Cardone, V. 2017. I reportages di viaggio per la conoscenza della città, in G. Belli, F. Capano, M.I. Pascariello (a cura di), La città, il viaggio, il turismo. Percezione, produzione e trasformazione. Napoli: Cirice, 2017, pp. 391–404.

Cervellini, F. 2017. Immagini di città, tra la scena, il labirinto e lo sprawl in G. Belli, F. Capano, M.I. Pascariello (a cura di), La città, il viaggio, il turismo. Percezione, produzione e trasformazione. Napoli: Cirice, 2017, pp. 413–416.

Chaillou, M. 1992. La mer, la route, la poussière, in AA.VV., Pour une littérature voyageuse. Bruxelles: Éditions Complexe, pp. 57–81.

Cometa, M. 1999. Il romanzo dell'architettura. La Sicilia e il Grand Tour nell'età di Goethe. Roma-Bari: Editori Laterza.

De Rubertis, R. 1994. Il disegno dell'architettura. Roma: La Nuova Italia Scientifica.

De Seta, C (a cura di). 2001. Grand Tour. Viaggi narrati e dipinti. Napoli: Electa Napoli.

De Seta, C. 1982. L'Italia nello specchio del "Grand Tour", in "Storia d'Italia", Annali, n. 5, Il Paesaggio (a cura di Cesare de Seta). Torino: Einaudi, pp. 127–263.

De Seta, C. 1999. Vedutisti e viaggiatori in Italia tra Settecento e Ottocento. Torino: Bollati Boringhieri.

Duque, F. 1994. La residenza dell'uomo superficiale, in Zarone (a cura di), 1994, pp. 45–53.

Gautier, T. 1863. Dessins gravés de Vctor Hugo – Texte par Théophile. Paris: Castel éditeur, n.e. Montpellier: L'Archange Minotaure (2002).

Georgel, P. 2002. Victor Hugo. Dessins. Paris: Gallimard.

Grau, C. 1973. Borges y la arquitectura. Madrid: Cattedra.

Hernández, M. 1998. Libro de los dibujos de Federico García Lorca. Granada: Editorial Comares - Fundación Federico García Lorca.

Herrmann Fiore, Kristina (a cura di). (2007) Dürer e l'Italia, catalogo della mostra omonima, Roma, Scuderie del Quirinale, 10 marzo – 10 giugno 2007. Milano: Electa.

Leed, E.J. 2001. Memoria e ricordo: il ruolo dei dipinti nel Grand Tour in Italia, in De Seta (a cura di), (2001), pp. 12–19.

Maglio, A. 2009. L'Arcadia è una terra straniera. Gli architetti tedeschi e il mito dell'Italia nell'Ottocento. Napoli: Clean.

Magris, C. 2005. L'infinito viaggiare. Milano: Mondatori.

Mangone, F. 2002. Viaggi a sud. Gli architetti nordici e l'Italia. Napoli: Electa Napoli.

Miller, S. 2001. Dibujos de Benito Pérez Galdós, Álbum arquitectónico. Las Palmas de Gran Canaria: Ediciones del Cabildo de Gran Canaria.

Moleón, P. 2003. Arquitectos españoles en la Roma del Grand Tour, 1746–1796. Madrid: Abada Editores.

Neumeister, S. 1994. La città come teatro della memoria. Argomenti letterari, in Zarone (a cura di), 1994, pp. 55–74.

Nuvolati, G. 2013. L'interpretazione dei luoghi. Flânerie come esperienza di vita. Firenze: Firenze University Press.

Pavia, Ro. 2015. Il passo della città. Temi per la metropoli futura. Roma: Donzelli Editore.

Pratelli, A. 1989. Viaggiare con un piccolo blocco da disegno. Strade itineranti della didattica e della storia del disegno d'architettura, in "Relazioni e Memorie" del xi Convegno Nazionale dei docenti delle discipline della rappresentazione nelle facoltà di architettura e di ingegneria, Lerici, 16, 17 e 18 ottobre 1989.

Reim, R. 2001. La Parigi di Zola. Roma: Editori Riuniti.

Schabert, T. 1994. La creatività e il nulla. Due esperienze dell'architettura della città, in Zarone (a cura di), 1994, pp. 11–26.

Villanueva, D. 2008. Imágenes de la ciudad. Poesía y cine, de Whitman a Lorca. Valladolid: Universidad de Valladolid.

Wunenburger, J-J. 1997. Philosophie des images. Paris: Presses Universitaires de France, 1997; trad. it. Filosofia delle immagini. Torino: Einaudi Editore, 1999.

Zarone, G. (a cura di). 1994. La città come destino dell'uomo. Esistenza spazio architettura. Napoli: ESI.

Drawing (…) City (…) Body, Dwelling on Earth – Janeiro (Ed.)
© 2019 Taylor & Francis Group, London, ISBN 978-0-367-02602-8

Conversions

E.W. Angelico
Department of Architecture, Polytechnic School, University of Palermo, Italy

ABSTRACT: Nowadays, several different forms of resilience draw our attention but, among all the inflated meanings of this term, we are definitely attracted by the one concerning the ability the material has to regenerate to a new life (in a different present) when it is widely used both in "sophisticated" technologies and also in "simple" methodologies to support man in his research, knowledge and production in a world still to be planned, far from the logic of reinforced concrete.

Today the anthropic world of "buildings" must be reconsidered in terms of redesign, systemic editing, methods and results of the performance while creating (even in the case of new artifacts).

The current environmental emergencies force us to adopt the principle of "Circular Economy" also in Architecture. It had never been considered as a complete cycle: at the time of its birth (project), its life (realisation and durability), its death (deconstruction and reuse). Dry stone architectures and other parallel techniques allow easy interventions with the purposes of "build on what has been built", "redesign", and "recycling". These actions constitute different resilient urban settings, able to reborn, to defend themselves and to activate again before adversities (earthquakes, natural disasters and anthropic emergencies).

Keywords: Technology, Resilience, Conversion, Reuse

"Resilience, the capacity to lead to a continued existence
by incorporating change"
(Berkes, Colding, Folke, 2003, p. 393)[1]

1 AT THE BEGINNING

The term 'resilience' comes from metallurgy: it represents the property of metal to resist the forces applied to it, in metallurgic technology.

For a metal, resilience is the opposite of fragility.

The same goes for psychology: a resilient person is opposed to an easily vulnerable one, which "lives: The NIGHT of ego" (Mormese, C. 2004). Etymologically, "resilience" comes from Latin "resalio", iterative of "salio". Some propose a suggestive connection between the original meaning of "resalio", implying the gesture of going back on an upturned ship capsized by the force of the sea, and its current use in the psychological field: both terms indicate the will to carry on without giving up while facing adversities, 'beyond' the night.

1. The use of ecosystem resilience together with the development of territorial systems has officially entered International and European Union policies since 2002, when Carl Folke, Steve Carpenter, Thomas Elmqvist et al., presented Resilience and Sustainable Development: Building Adaptive Capacity in a World of Transformations, and today has a central role in community policies (EAC, 2002).

Figure 1. The strength to get a new lease on life, feeling alive after experiencing the worst.

2 ON THE DEFINITION OF PIETRO TRABUCCHI

Psychological resilience is the ability to persist while achieving challenging goals, effectively coping with difficulties and other negative events that will be on our path. The verb "to persist" indicates the idea of a solid, unwavering motivation.

In fact, the resilient individual has a series of unmistakable psychological characteristics: he/she is optimistic and tends to 'see' negative events as temporary and limited; he/she believes having a wide degree of control on his/her life as well as on the surrounding environment; he/she is strongly driven to achieve pre-established goals; he/she views change as a challenge, an opportunity, and not as a threat; when faced with defeat and frustrations he/she doesn't lose hope.

3 CURRENT USE

'Resilience' is the ability to positively cope with traumatic events, successfully reorganise one's life while experiencing difficulties. It is the ability to regain one's life by remaining sensitive to the positive opportunities that life offers, without losing humanity.

Resilient **people**, surrounded by adverse circumstances, despite all the obstacles and sometimes against all odds, manage to effectively face setbacks, to give new impetus to their lives and even to reach important goals. It doesn't mean to be strong, but flexible, therefore manage to sensibly bend to events, and react accordingly (The road of resilient – American Psychological Association (APA) – official site).

Resilient **territories**, exist in a concrete geographical dimension, fragmented into many pieces by wars, conflicts, catastrophes and cataclysms.

In these places, despite open wounds, there is an innate 'resilience' of recreation, an amazing ability to find in itself the means for a new phase of growth, affirmation, self-determination for regeneration. The strong presence of natural elements, allows a high resilience: the trees resist, their roots' net contains avalanches and drought, the plants regenerate, the animals repopulate the area. Life recovers even after a strong stress, traumatic or disastrous events. Everything regenerates.

For us, these subjects represent important experimentation and research fields, on which we should focus to understand the actions that are set in motion, both spontaneously and through the effect of sensible human choices. Consequently, codify the events to draw pilot actions and activate new research areas.

Resilient cities, generally are artificial places, completely built, mostly man-made, and have a 'low' ability to resist the environmental shocks (heat waves, floods, extreme weather events, earthquakes) and to recover from damages.

Figure 2. Goqui Island (China), has been claimed by nature in the most stunning way.

After a hurricane or a fire, the artificial places are damaged beyond the possibility of recovery.

This means, they have a low, insufficient or even null resilience. In these conditions, there is very little to do and it is possible to rely on specific energies (especially economic and forecasting types): forward projections that anticipate deteriorations and events.

4 WHAT ABOUT ARCHITECTURE?

In accordance with the above-mentioned considerations, it might seem that Architecture is 'passive'. Technically, it is the less resilient setting, less susceptible to autonomous rebirth. That is why, we need to ask ourselves how we might consider it particularly resilient, and what actions we should implement to make it resilient and understand how to connect natural and artificial factors, carefully looking for the priority of one or the other.

On the one hand, the constituent materials of architecture can contribute (they can certainly resist more in 'endurance' and 'durability'). On the other hand, the project (ab-origine, from its creation) with resistance properties of its construction systems and its acclaimed static benefits. Furthermore, rethinking architecture as built by modules, so that it does not go into pieces since it was originally built with pieces.

It is a clear reference to the well-known dry stone construction systems (to be assembled) – definitely much more solid and resilient of others—that by being built with modules can be easily, quickly and effectively be replaced when necessary. Such systems, appear to be much more elastic and flexible when an earthquake occurs or in case of hydro geological instability. Therefore, they are more resilient than ordinary construction systems (reinforced concrete and similar).

Of course, the subject is much more complex than that, and we should, first of all, consider it as a historic-cultural matter. The Architecture has always been imagined as 'long-lasting matter', presumptuously capable of challenging time with its mass and its size (ancient Greek or Roman heritage). However, current history has shown otherwise.

Throughout the evolution of buildings, man has always laid the foundations for substantial achievements, aiming to persist and presumptuously remain unharmed in time. This feeling increased with the creation of reinforced concrete (E. Mendelson's Einsteinturm Tower, 1917) considered the perfect means for human creativity and capable of embrace all his wills.

Architecture has indelibly affected the territory (instead of consuming it) and its transformations have never gone unnoticed, even after decades. Wastes of disused buildings float in the confused nebulae of our suburbs and man does not know how to contain the potential

Figure 3. Detroit (USA), "Motor City" ford headquarters where more than three million "T" models were produced.

economic bleeding if he wanted to recover, dismantle or demolish them: bottomless processes that will never find a solution (Detroit – USA)[2].

Bad building (Bottone, P. 2015), together with territory and construction site misgovernment, have already shown signs of dismantling for decades and its results are before everyone's eyes.

We need to drastically abandon planning interventions if competence, actions designed for the context and evaluation of side effects are lacking. Therefore, stop designing on a big scale and creating a timely design, and, never again, consider a building as a 'sacrifice' for knowledge or creativity (whose results are often nefarious and irrecoverable).

Sadly, we have too long been used to consider the 'error variable' as regular requirements of projects and buildings. We easily accept the 'error variable' as standard action; errors in the design, determination of quantities, prices, necessities, needs, supply, logistics; errors in the installation of materials, times and deliveries; errors considered in the call for tenders under the heading, "risk of error"; and even the errors arising from economic assessments throughout the life cycle of the building (Argiolas, C., Quaquero, E., Prenza, R., 2015).

Therefore, we have accepted the error as part of the making process and, its complete incapability of being 'resilient'. Without thinking about the consequences of such errors that affect not only the life of the building, but also its scope, management and comfort.

Today, stand against this logic is a true necessity, a duty, a goal for the future. Think about these subjects fits the logic of "misjudge/correct". Technological development leads us to eliminate the superfluous, lighten, eliminate matter and, instead, add thought, intelligence, "suitable" creativity; know and avoid the mistakes that others have made before us while trying to keep up with times, to which we must necessarily adapt (Zennaro, P., 2000). Materials, techniques, knowledge, lifestyles, climate and general innovation undergo continuous muta-

2. Detroit – "What do I have to do to be saved?" The huge writing on the building next to Michigan Central Station well expresses the sense of dismay and resignation of Detroit's soul. Fifty years ago: the District's black community exploded with rage, the city burned, the US military brought order, and the gap with white people became a deep wound. 1970s and 1980s: the city emptied (only a quarter of the 2 million citizens of the 50s) and it became a "Ghost Town", one of the most violent in the United States. 1990s: globalisation damaged the foundations of Detroit Motor City. 2000s: the car industry went out of business because of Asian competition. In July 2013, the town council, burdened by a debt of 18 billion dollars, declared bankruptcy (from Il Sole 24 Ore on-line).

Figure 4. Casa Sabugo, Gijón, Spain (2016), Arch. Tagarro; Arch. De Miguel: Single-family homes restoration.

Figure 5. Wikkelhouse (2017), developed by the Dutch company fiction factory, Is made out of cardboard, modular and highly customisable, built to last between 50 and 100 years.

tions and it will be difficult to find a coherent solution for different problems: however, the effort must be made!

To build means: lift up by the union of conveniently placed parts.

We believe the most coherent property is established by environmental sustainability, reuse, 'recycling' and the circular economies originating from the "do/misjudge and correct" process, to reintroduce the efforts made in new circular processes (feedback) that can consume only what they produce.

Therefore an Architecture produced by mechanics rather than masons; an Architecture capable of reconverting with little effort, precisely because created ab-origin with interchangeable modules.

5 CONCLUSIONS

As a Phoenix, mythological animal, symbol of natural and evolutionary cycles of 'death and rebirth' – reason for the evolution, life and 'resilience' of men and nature—Architecture is to be considered able to respond to this process, capable of reconvert at any time into 'something else', as a different already foreseen reality (Carta, M. 2014). We are certain that only an Architecture in pieces and made by modules can truly and positively fulfil this new need.

As it happens in nature (see Mimosas or Bamboo)[3] the elements regenerate and autonomously strive to take back their space. We saw that even a healthy man is capable of regaining his life through positive attitudes of rebirth. Architecture must have the same ability of resilience, to restart and re-establish its building processes.

We have to take note that every building (added) on territory must only 'occupy' the territory it stands on without 'consuming' it. Therefore the new buildings must be designed keeping in mind three key moments:

- birth (creativity, harmony, research, experience, context, technology, forecasting, solutions);
- life (maintenance, social and environmental conditions, transformation, conversion, adjustment, immediate reinvention, adaptation, global or partial redevelopment, resilience to events, expansion/reduction of space);
- death (partial or total deconstruction; conversion and reconversion, partition, reuse, recycling, reintroduce pieces into cycles, readaptation of parts, regeneration of waste).

This will set a new way of thinking about the buildings, where error can become a research tool, having the ability to self-correct in every moment.

REFERENCES

Argiolas, C., Quaquero, E., Prenza, R., 2015. from Dal disegno alla simulazione: Nuovo paradigma per il progetto e la produzione edilizia; Gangemi Editore spa, ISBN 978-88-492-9539-9.

Berkes, F.J.F., C.. 2003. From Navigating social-ecological systems: Building resilience for complexity and change. Cambridge University Press, Cambridge, UK. 393 pp., ISBN: 0-5218-1592-4.

Bottone, P., 2015. from Dark Italy: Cronache dal lato oscuro del belpaese; edited by, Edizioni Sette Città, ISBN 978-88-7853-377-6.

Carta M., Lino B., Ronsivalle D. 2016. (edited by), from Re-cyclical Urbanism. Visioni, paradigmi eprogetti per la metamorfosi circolare, Trento-Barcelona, ListLab, ISBN 889985422X, 9788899854225.

Carta, M., 2014. from "Re-imagining the city: Progettare il nuovo metabolismo urbano", in Zazzero E. EcoQuartieri. Temi per il progetto urbano sostenibile, Sant'Arcangelo di Romagna, Maggio.

Castelli, C., 2011. from "Resilienza e creatività. Teorie e tecniche nei contesti di vulnerabilità", Published by Franco Angeli, ISBN 8856833735, 9788856833737.

Mormese, C. 2004. from "Strega: ombre di libertà", Lampi di Stampa, p. 91/96, ISBN 88-488-02621.

Trabucchi, P. 2007. from "Resisto dunque sono" – Published by Corbaccio,, ISBN 887972911X, 9788879729116.

Treccani, G. 2017. from Costruire: Nuovo Vocabolario Treccani, – Istituto della Enciclopedia Italiana—Published by Giovanni Treccani S.p.A.

Zennaro, P., 2000. From "La qualità rarefatta: Considerazioni sull'influenza del vuoto nella costruzione dell'architettura; – Published by Franco Angeli, ISBN 88-464-2556-1.

3. In 1946 the U.D.I (Unione Donne Italiane) chose the Mimosa as perfect flower to represent women's day. In fact, it is a flower that grows spontaneously in many parts of Italy. Only a sprig of Mimosa was spared by the fire of a textile factory where dozens of women died, so it was chosen to honour their memory (resilient).

Drawing (...) City (...) Body, Dwelling on Earth – Janeiro (Ed.)
© 2019 Taylor & Francis Group, London, ISBN 978-0-367-02602-8

Author index

Abdalla, C.G. 101
Angelico, E.W. 213

Baptista-Bastos, M. 147
Barbosa, A. 1
Bechmann, I. 87

Cardone, V. 203
Carvalho, G. 87
Catalfamo, V. 137
Costa, M. 111
Cruz, M. 27

da Luz Moreira, P. 163
D'Alba, V. 65
D'Auria, S. 203
De Feo, E. 203
Duque, R. 1

Fatta, F. 51
Ferreira Crespo, J. 95
Ferreira Lopes, R. 179
Figueiredo, M. 87

Giunta, C. 189
Godinho, A. 7

Janeiro, P. 173
Jorge, L. 121

Khoury, F. 155

Lima, F. 37
Loução, D. 27
Louro, M. 137

Maggiore, F. 65
Marraffa, A. 51
Mendes de Vasconcellos, L. 179
Milheiro, A. 15
Moreira da Silva, A. 21
Moreira da Silva, F. 45
Moschini, F. 65

Nascimento, M. 155

Oliveira, F. 79

Rocha, J. 111

Santos, C. 101
Silva, G. 37
Sol, L. 131